Scottish Heinemann Maths

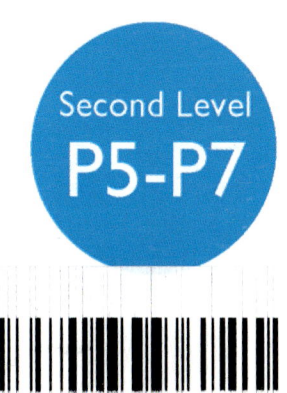

Second Level
P5-P7

D1799864

Delivering the Curriculum for Excellence

Heinemann is an imprint of Pearson Education Limited, a company incorporated in England and Wales, having its registered office at Edinburgh Gate, Harlow, Essex, CM20 2JE. Registered company number: 872828

www.heinemann.co.uk

Heinemann is a registered trademark of Pearson Education Limited

Text © Scottish Primary Mathematics Group 2008

First published 2008

12 11 10 09
10 9 8 7 6 5 4 3 2

British Library Cataloguing in Publication Data is available from the British Library on request.

ISBN 978 0 435 17127 8

Designed and typeset by TechType
Illustrated by Gecko Limited, Bicester, Oxon and Oxford Illustrators
Cover design by Christopher Howson
Cover photo/illustration © Pearson Education Ltd/Jules Selmes
Printed in the UK by Ashford Colour Press Ltd

Acknowledgements
We would like to thank the head teacher, teaching staff and pupils at Jordanhill Primary School, Glasgow for their invaluable help in creating the video and photographs on the CD-ROM.

Every effort has been made to contact copyright holders of material reproduced in this book. Any omissions will be rectified in subsequent printings if notice is given to the publishers.

Contents

Publisher's Foreword

This book is a revised and expanded version of the *Organising and Planning Guides* for *Scottish Heinemann Maths* with additional referencing to the Numeracy and Mathematics Outcomes of A Curriculum for Excellence. It is produced by SPMG and Heinemann with the assistance of an author who was formerly a key member of the maths team at Learning and Teaching Scotland.

As well as an introduction to the components of *Scottish Heinemann Maths* and an explanation of the pedagogy behind this popular programme, this book also contains charts and tables to support you in planning, teaching and assessing your class. All of the relevant charts are also provided on the accompanying CD-ROM, along with some video clips and photographs exemplifying how users of *Scottish Heinemann Maths* around the country are beginning to implement A Curriculum for Excellence.

The Curriculum for Excellence offers a relevant, unified curriculum which will prepare our young people with a broad range of skills and experiences for an ever changing and unknown future. It outlines the experiences our young people should have to enable them to become successful learners, confident individuals, responsible citizens and effective contributors. It focuses on how we teach rather than what we teach by highlighting the principles of challenge and enjoyment, breadth, depth, progression, personalisation and choice, coherence and relevance.

Implementing A Curriculum for Excellence may mean you need support with looking again at your maths planning and assessment. The levels have been defined in a broader manner than in the 5–14 guidance with Early level bridging the transition from nursery to P1, First level covering P2 to P4 and Second level covering P5 to P7 for most pupils. The progression across the levels is not content specific as previously. It doesn't specify, for example, the range of numbers or types of shapes. It is open for teachers to use their ongoing formative and summative assessments to decide where a pupil is at and how to move them forward. The outcomes focus on the progression of wider concepts, such as moving from sharing and grouping to more complex mental and written skills of multiplication and division.

A Curriculum for Excellence has been designed to allow you to use your professional judgement to evaluate the content, contexts and methodologies which are most appropriate for your pupils. We hope that the advice and resources provided herein help you to plan the right progression of content for your pupils using *Scottish Heinemann Maths*. For the most up-to-date support, please log on to our website, **www.heinemann.co.uk**.

Note on the Development Planners (pages 40–69)

As Curriculum for Excellence outcomes covered within *Scottish Heinemann Maths* are arranged into Early, First and Second levels rather than Primary 1, Primary 2 and so on, each Development Planner covers outcomes from more than one level. The codes for the outcomes indicate which level each outcome is from, for example:

┌─ The **1** indicates First level   ┌─ The **2** indicates Second level

MTH 114N   **MNU 232W**

Introducing Scottish Heinemann Maths

Mathematical development

Scottish Heinemann Maths is a course that facilitates implementation of the guiding principles and the approaches to learning and teaching described in Curriculum for Excellence. It can also help teachers to ensure that their pupils achieve the specific Curriculum for Excellence Outcomes in both Numeracy and Mathematics and, in so doing, experience challenge and enjoyment. Its clearly defined structure provides progression from Primary 1 to Primary 7 and offers schools a teaching programme with both coherence and continuity.

There is an emphasis on direct, interactive teaching aimed at helping children to develop understanding and confidence in relation to a range of mathematical concepts, skills and processes. These include the ability to recall basic facts quickly, employ, with confidence, a variety of mental strategies and written methods of calculation, use appropriate mathematical vocabulary and make connections between different topics in mathematics, with other areas of the curriculum and children's own everyday experiences.

Effective teaching

Each mathematical topic in the *Teaching Files* is developed systematically through a series of carefully structured lessons and linked *Pupil Activities*. The lessons provide opportunities for active teaching methodologies, and oral and mental work involving groups or the whole class on a daily basis.

Pupil Activities are provided to consolidate key teaching ideas and support group work, differentiation, discussion and pupil involvement. Many of the small group activities require the children to work collaboratively, for example, in order to play a game or investigate a problem. Suggestions on a range of simple and effective teaching resources to help motivate the children, illustrate key teaching approaches, or enhance participation, are included in each section of the *Teaching Files*.

Classroom organisation

The components of **Scottish Heinemann Maths** are designed for use in a flexible way. This ensures that the needs of children and teacher are met, whatever form of classroom organisation is used. Interactive *Teaching* activities can be used, as appropriate, with groups or a whole class. There are suitably differentiated *Pupil Activities* and written practice, consolidation, application and extension work is provided in the *Textbooks*, *Activity Books* and *Extension Textbooks*. These are also designed to accommodate use with groups, individuals or the whole class.

Mental calculation

The course stresses the importance of children developing the ability to 'work things out in their heads'. There is, therefore, an emphasis on oral, mental mathematics, rapid random recall of basic number facts and children justifying and explaining their methods. To acquire the necessary skills and confidence to do this, number facts and a range of mental calculation strategies are taught and practised in a progressive and systematic way.

Developing mathematical thinking

There is an emphasis throughout on investigative approaches to teaching and learning that encourage children to ask questions and explore alternatives as a means of deepening their understanding and developing their appreciation of the purposes and value of the mathematics they have learned. Problem solving strategies are introduced in a sequential way that promotes collaboration and discussion.

Planning for learning

Detailed advice and examples of long- and short-term plans are given on pages 19–21 of this book. The emphasis is on creating a coherent and manageable form of planning that complements school or education authority guidance.

Assessment and recording

Scottish Heinemann Maths provides a range of assessment materials to support teachers in implementing the recommendations of *Assessment is for Learning*. They are designed to enhance pupil learning by helping the teacher build up a detailed picture of the children's attainment and to check that anticipated outcomes are being achieved. Advice and assistance is also given to support the teacher's ongoing formative assessment, which is carried out on a daily basis by interacting with children or observing them at work.

The assessment materials may cover a short section of work in a *Check-up*, a whole topic in a *Topic Assessment*, or provide an end-of-year *Round-up* where different areas of mathematics are assessed. The materials can also be used as part of the important process of giving regular feedback to children that allows them to see the progress they have made. They also provide a comprehensive record of achievement that can be shared with the children, their parents and other teachers.

Involving parents

Home Activities can be used to support a school's commitment to actively involving parents in their children's learning. They provide a number of straightforward activities that give parents confidence in helping their child with mathematics. For the child, *Home Activities* provide opportunities for further practice and consolidation.

For teachers: *Organising and Planning Guides*

Teaching Files

Teaching Resource Books

Answer Books

For children: *Textbooks*

Extension Textbooks

Assessment Books (including *Check-ups*, *Topic Assessment* and *Round-up tests*)

Pupil Sheets (included in the *Teaching Resource Books*)

Home Activities (included in the *Teaching Resource Books*)

Resource Sheets (included in the *Teaching Resource Books*)

Pupil Software

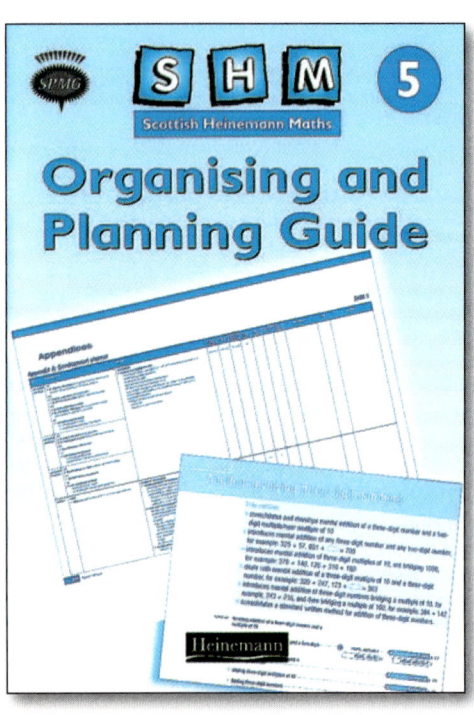

Organising and Planning Guides

- The guides outline:
  - the main features of the course
  - the component parts of **SHM 5**, **SHM 6** and **SHM 7**.
- They provide advice about:
  - planning to use the course effectively
  - organising resources
  - teaching lessons and follow-up work
  - assessing learning.
- Also included are:
  - charts to show the mathematical content of **SHM 5**, **SHM 6** and **SHM 7**
  - a mapping to Level C, Level D and Level E attainment targets in development planners for the year
  - an example of a weekly planner
  - pupil record charts
  - an assessment record grid
  - Level C, Level D and Level E class record grids.

Teaching Files

- The files contain:
  - a bank of *Starters and other mental activities* to supplement those included in lessons throughout the file
  - teaching notes giving suggestions for lessons, pupil activities, further teaching, the use of the *Textbooks* and *Extension Textbooks* pages, follow-up activities and assessment
  - references to photocopiable *Pupil Sheets* for use either within a lesson or as follow-up practice, consolidation or extension
  - references to photocopiable *Home Activities* to give homework linked to the work in school.

Starters and other mental activities

- The *Teaching Files* have a bank of suggestions for oral mental activities. These are intended to promote a 'feel' for number, quick recall of number facts and the flexible use of mental calculation strategies. They are designed to be interactive, involving the teacher and a large group or the whole class.

- Used as 'starters' at the beginning of lessons, the activities help to keep skills 'ticking over', even when the main teaching has moved to another topic. They can, however, be used at any time for practice or consolidation, as well as to check whether children are ready for the next step.

- Some of the activities are *generic*, providing a 'format' which can be adapted to suit different topics. The remainder are linked to *specific* topics.

Teaching notes

- **SHM 5** includes the following mathematical topics:

Numbers to 100 thousands	Division	Weight
Time	Addition	Fractions
Decimals	3D shape	Subtraction
Money	Area	2D shape
Addition and Subtraction beyond 1000	Number properties	Volume and capacity
Position, movement and angle	Multiplication	Length
Data handling		

- **SHM 6** includes the following mathematical topics

Numbers to millions	Number properties	Length
3D shape	Addition	Fractions
Weight	Position, movement and angle	Subtraction
Decimals	Volume/Capacity	Data handling
Multiplication	Percentages	Area
Division	Time	2D shape

- **SHM 7** includes the following mathematical topics

Place Value	Number properties	Weight
3D shape	Addition	Fractions
Length	2D shape	Subtraction
Decimals	Time	Position, movement and angle
Multiplication	Percentages	Volume/Capacity
Data handling	Division	Problem solving and enquiry
Area		

- At the beginning of each topic a summary page provides:
  - a description of related *Previous work*
  - a concise *Overview* of the work of the new topic
  - a *Development* section, which details the mathematical content and suggested teaching approaches for the topic

Division

Previous work

In *SHM4* the work on *Division* extended mental division by 2, 3, 4, 5 and 10 to mental division by 6, 7, 8 and 9, consolidated linking multiplication and division, dealt with remainders and rounding answers in context, and introduced division of two-digit numbers beyond the tables using informal and standard written methods.

Overview

The work on *Division*:
- revises mental division by 2–10
- includes divisibility tests for the divisors 2, 4, 5, 9 and 10
- consolidates linking multiplication and division including the link between doubling and halving
- introduces division of a three-digit number by a single digit using informal and standard written methods
- consolidates dealing with remainders and rounding answers in context.

Development

- Knowledge of the multiplication tables for 2–10 is used to consolidate mental division by 2–10, for example: $48 \div 6$ can be found by asking *Six times what is forty-eight?* or *How many sixes are in forty-eight?* The ability to recall division facts is developed by giving practice in rapid mental calculation involving random division by 2–10.
  The 'fractional' form of recording division, for example $\frac{63}{7}$, is introduced as an alternative to the established $63 \div 7$.
- Awareness of digit shift patterns in division by 10 and by 100 of
  - a three-digit multiple of 100 ($800 \div 10$, $800 \div 100$)
  - a four-digit multiple of 1000 ($7000 \div 10$, $7000 \div 100$)

  is extended to
  - division by 10 and by 100 of a four-digit multiple of 100 ($4200 \div 10$, $4200 \div 100$)
  - division by 1000 of a four-digit multiple of 1000 ($5000 \div 1000$).
- The idea of a divisibility test is introduced by investigating exact divisibility by 9 (the sum of the digits of the dividend is divisible by 9). Thereafter the children explore divisibility tests for the divisors 2, 4, 5, 10 and 100 which are based on inspection of the final one or two digits (for example, a number is divisible by 5 if its final digit is 0 or 5).
- The relationship between doubling and halving, which is consolidated then extended, includes finding halves of
  - even numbers to 200 ($\frac{1}{2}$ of 168)
  - any three-digit even numbers ($\frac{1}{2}$ of 692)
  - 'even' multiples of 10 to 2000 ($\frac{1}{2}$ of 1360, but not $\frac{1}{2}$ of 1370)
  - 'even' multiples of 100 to 10 000 ($\frac{1}{2}$ of 5600, but not $\frac{1}{2}$ of 5700)

  The main strategy used involves partitioning, for example, for $\frac{1}{2}$ of 168:
  Half of 100 is 50. Half of 68 is 34.
  So, half of 168 is 50 plus 34 (84).

174 Division

- a *Contents* table containing references to teacher and pupil materials required for each section within the topic
- a *Language* list of relevant mathematical vocabulary
- a *Resources* list, which outlines general materials and specific *Resource Sheets*.

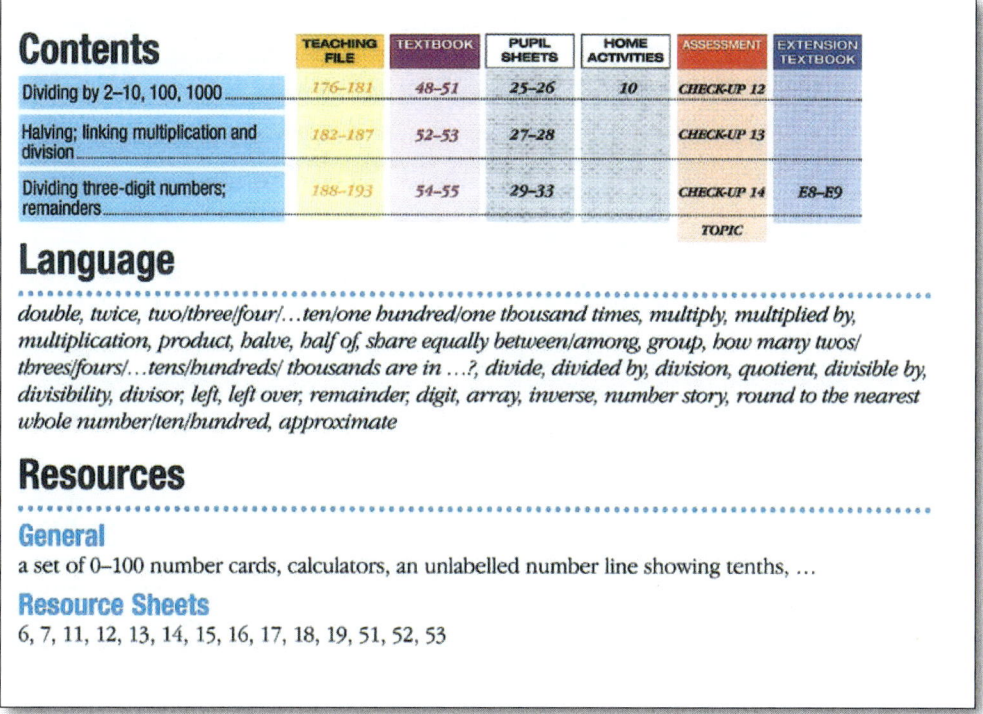

Contents

	TEACHING FILE	TEXTBOOK	PUPIL SHEETS	HOME ACTIVITIES	ASSESSMENT	EXTENSION TEXTBOOK
Dividing by 2–10, 100, 1000	*176–181*	*48–51*	*25–26*	*10*	*CHECK-UP 12*	
Halving; linking multiplication and division	*182–187*	*52–53*	*27–28*		*CHECK-UP 13*	
Dividing three-digit numbers; remainders	*188–193*	*54–55*	*29–33*		*CHECK-UP 14*	*E8–E9*
					TOPIC	

Language

double, twice, two/three/four/…ten/one hundred/one thousand times, multiply, multiplied by, multiplication, product, halve, half of, share equally between/among, group, how many twos/threes/fours/…tens/hundreds/ thousands are in …?, divide, divided by, division, quotient, divisible by, divisibility, divisor, left, left over, remainder, digit, array, inverse, number story, round to the nearest whole number/ten/hundred, approximate

Resources

General

a set of 0–100 number cards, calculators, an unlabelled number line showing tenths, …

Resource Sheets

6, 7, 11, 12, 13, 14, 15, 16, 17, 18, 19, 51, 52, 53

- The notes for each section within a topic follow the same pattern.
  - a brief statement outlines the work covered by the section.
  - a *Schematic* diagram details the lessons within the section. It shows how all the associated materials in **SHM** fit together and progress through the work of the section.

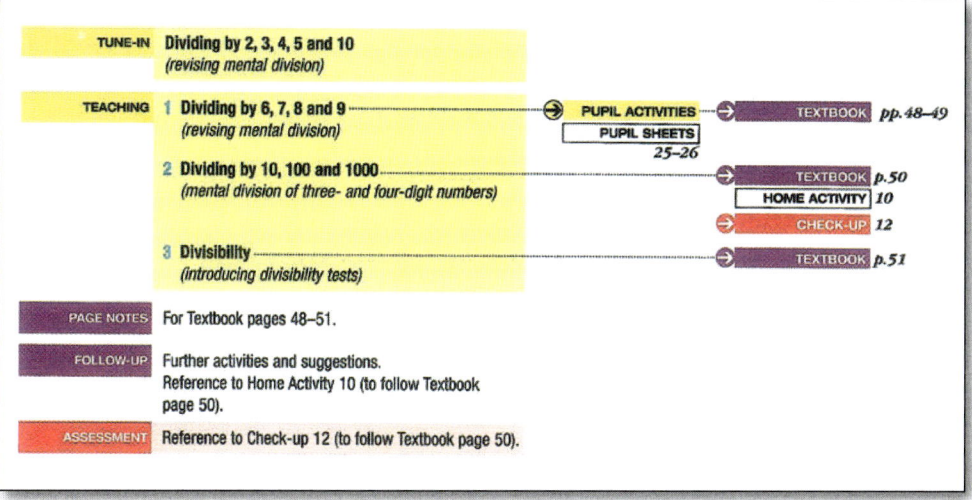

TUNE-IN	**Dividing by 2, 3, 4, 5 and 10** *(revising mental division)*	
TEACHING	1 **Dividing by 6, 7, 8 and 9** *(revising mental division)*	PUPIL ACTIVITIES → TEXTBOOK *pp.48–49*; PUPIL SHEETS *25–26*
	2 **Dividing by 10, 100 and 1000** *(mental division of three- and four-digit numbers)*	TEXTBOOK *p.50*; HOME ACTIVITY *10*; CHECK-UP *12*
	3 **Divisibility** *(introducing divisibility tests)*	TEXTBOOK *p.51*
PAGE NOTES	For Textbook pages 48–51.	
FOLLOW-UP	Further activities and suggestions. Reference to Home Activity 10 (to follow Textbook page 50).	
ASSESSMENT	Reference to Check-up 12 (to follow Textbook page 50).	

TUNE-IN

A *Tune-in* is then provided as a suggestion for starting the teaching. This is an interactive, mental, whole-class activity which revises any relevant previous work and sets the scene for the first lesson of the section.

An activity from the bank of *Starters and other mental activities* can be used, if required, as a lead-in to subsequent lessons. Alternatively, an activity from the previous day's work can be adapted for this purpose.

TEACHING

Teaching suggestions are given for group or class lessons to develop the sequence of work in the section.

PUPIL ACTIVITIES

Pupil Activities, including practical activities, games and mental work, follow many of the lessons. These are designed to be used by groups, pairs or individuals, with some teacher support. When several activities are provided, a selection should be made to suit groups within the class.

References to *Textbook* pages point to appropriate written work for pupils.

 **ACTIVITY BOOK**
Page 10

 **TEXTBOOK**
Page 26

 **FURTHER TEACHING**
1

FURTHER TEACHING

References are given to *Further Teaching*, if any.

Suggestions for *Further Teaching* lessons with the class or a group provide an alternative approach, consolidation or extension.

PAGE NOTES

Page Notes offer advice about using some of the relevant *Textbook* pages. The notes highlight more challenging examples and possible difficulties with language and instructions.

FOLLOW-UP

Follow-up suggestions are given for drawing lessons to a close. These often involve discussion, mental work, extension activities or further practice.

References are also given to:

– *Home Activities* related to the *Textbook* pages.

ASSESSMENT

The *Assessment* section may include:

– a reference to a *Check-up* associated with the work of the section

– page notes for *Topic Assessments*.

The *Topic Assessments* are designed to assess for **SHM 5**: a number, money or time topic, covering the work of several sections, for **SHM 6** and **SHM 7**: a number, measure, shape or data handling topic, covering the work of several sections or a single section in the case of a short topic. The notes point out common errors that may occur and make some suggestions for dealing with them. For each question there are references to relevant *Textbook* pages and the appropriate section of the *Teaching File*, should some re-teaching or additional practice be considered necessary.

Teaching Resource Books

- The *Teaching Resource Books* contain:
  - photocopiable *Pupil Sheets* for use either within a lesson or as follow-up practice, consolidation or extension
  - photocopiable *Home Activities* to give homework linked to the work in school
  - photocopiable *Resources Sheets* for use by the teacher during lessons.

Pupil Sheets

- The **SHM 5** *Teaching Resource Book* includes 74 photocopiable *Pupil Sheets*. The **SHM 6** *Teaching Resource Book* includes 58 photocopiable *Pupil Sheets* and the **SHM 7** *Teaching Resource Book* includes 40 photocopiable *Pupil Sheets*.

- There are several types which have different purposes. For example:
  - to provide a means of recording during the course of a lesson
  - to give extra practice to children who have completed the *Textbook* pages but need more examples
  - to provide a template for teachers to produce their own sheets, which can be customised to cater for different ability levels.

Home Activities

- The **SHM 5** *Teaching Resource Book* contains 18 *Home Activities* and 4 *Home Sheets* of associated 'cards' for use with some of them. The **SHM 6** *Teaching Resource Book* contains 23 *Home Activities* and 3 *Home Sheets* and the **SHM 7** *Teaching Resource Book* contains 17 *Home Activities* and 3 *Home Sheets*. These are all photocopiable.

- The activities aim to:
  - provide important extra practice for the child
  - give parents an opportunity to be actively involved with their children's learning and provide encouragement and help
  - inform those at home about the mathematics being taught in school.

- *Home Activities* are referenced from:
  - the foot of the *Textbook* page which completes this work in school
  - the *Follow-up* section in the *Teaching File* for the related *Textbook* page.

- There are two types of *Home Activity*. (Often both types appear on one sheet. However, it is not necessary to use both parts at the same time.)

  - simple oral, mental activities or games involving an adult and the child (There are straight-forward instructions for the adult which give examples of the language to be used.)

  - written practice examples for the child to complete and the adult to check.

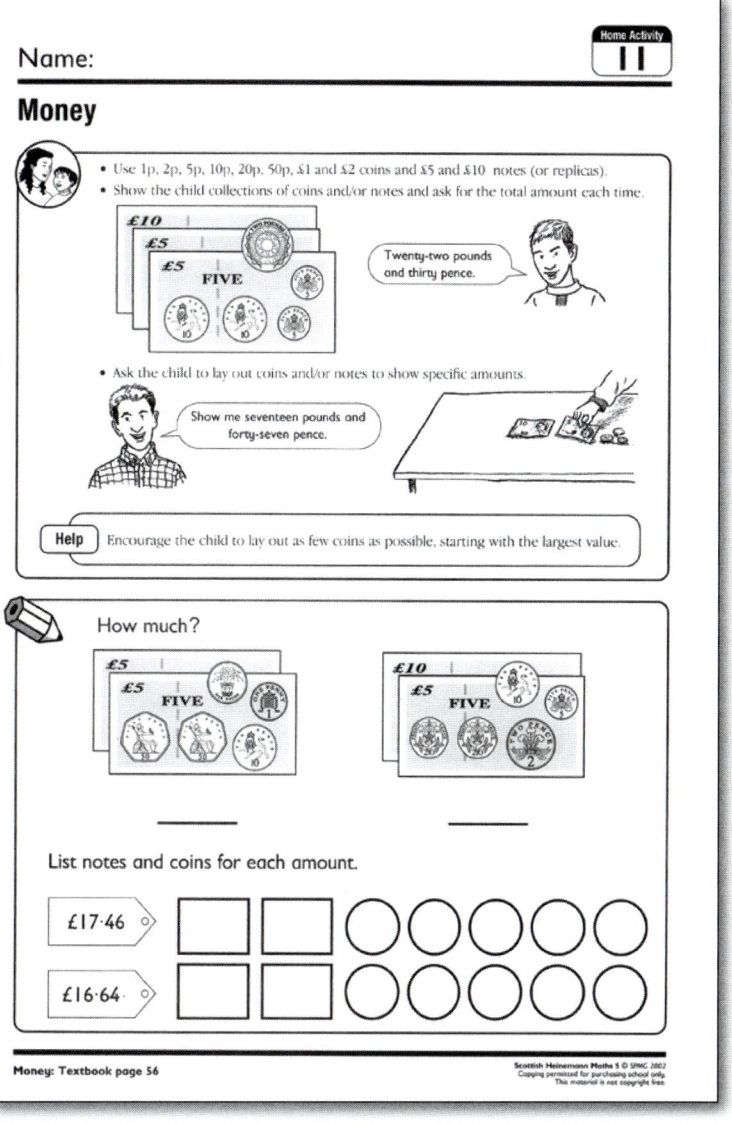

- The **SHM 5** *Teaching Resource Book* includes 54 photocopiable *Resource Sheets* for producing materials such as flashcards, number cards, 100-squares and so on, for use when teaching lessons. The **SHM 6** *Teaching Resource Book* includes 46 photocopiable *Resource Sheets* and the **SHM 7** *Teaching Resource Book* includes 30 photocopiable *Resource Sheets*.

Textbooks

- The *Textbooks* contain written work for the children for use after the *Teaching* of a lesson and related *Pupil Activities* have been completed. The pages provide practice, consolidation and application and can also be used as the basis for formative or pupil self-assessment.

- There are references at the foot of some of the *Textbook* pages to:
  - *Check-ups*, which assess one or two sections of work
  - *Home Activities*, which provide related work for a child and adult at home
  - *Topic Assessments*, which assess the work related to a specific number or money topic.

Extension Textbooks

- The *Extension Textbooks* include a range of activities to provide both lateral and vertical extension. Many of the addition, subtraction, multiplication and money activities involve the use of a calculator. The activities include mixed operations, 'multiple step' operations and mixed measures. There is encouragement throughout for children to discuss and collaborate when using and applying their knowledge and skills through problem solving and enquiry contexts.

Assessment Books

- The *Assessment Books* include:
  - *Check-ups*
  - *Topic Assessments*
  - *Round-ups.*

- They are designed to assess children's understanding, knowledge and ability to apply skills and techniques. The assessments are provided in both booklet and photocopy master format.

- The **SHM 5** *Assessment* book contains:
  - 21 *Check-ups* covering the topics shown:

Numbers in 100 thousands	4	Division	3
Addition	2	Fractions	1
Subtraction	2	Money	2
Multiplication	3	Decimals	2
		Time	2

  - 9 *Topic Assessments* covering:

Numbers in 100 thousands	Addition and Subtraction	Money
Addition	beyond 1000	Fractions
Subtraction	Multiplication	Decimals
Division		

- 3 *Round-up* tests, including a Level C test, containing questions on number, money, measure, shape and data handling.

- The **SHM 6** *Assessment Book* contains:
  - 19 *Check-ups* covering the topics shown:

Numbers to millions	3	Division	2
Addition	2	Fractions	2
Subtraction	2	Decimals	6
Multiplication	2		

  - 15 *Topic Assessments* covering:

Addition	Number properties	Time	2D shape
Subtraction	Fractions	Length	Position, movement and angle
Multiplication	Decimals	Weight	
Division	Percentages	Area	Data handling

  - 1 Level D *Round-up* test containing questions on number, money, shape, measure and data handling.

- The **SHM 7** *Assessment Book* contains:
  - 9 *Check-Ups* covering the topics shown:

Addition	1	Fractions	2
Subtraction	1	Decimals	5

  - 17 *Topic Assessments* covering:

Place value	Number properties	Weight	2D shape
Addition	Fractions	Length	Position, movement and angle
Subtraction	Decimals	Time	
Multiplication	Percentages	Volume	Data handling
Division	Area		

  - 1 Level E *Round-up* test containing questions on number, money, shape, measure and data handling.

- Each *Check-up* covers a smaller range of work within a single topic and is linked to the work of several *Textbook* pages. References to *Check-ups* are given in the *Teaching File* and at the foot of appropriate *Textbook* pages.

- The *Topic Assessments* cover work related to a whole topic. References to *Topic Assessments* are given in the *Teaching File*, at the end of the appropriate section of notes, and at the foot of the appropriate *Textbook* pages.

- The *Round-ups* cover a wide range of 'mixed' mathematics and give an indication of overall level of attainment.

Answer Books

- The *Answer Books* contain answers for the:
  - *Textbooks*
  - *Extension Textbooks*
  - *Pupil Sheets*
  - *Assessments*
  - *Home Activities.*

Delivering the Curriculum for Excellence

SHM Pupil Software

Each CD-ROM includes 60 engaging, differentiated pupil activities designed for use either on an interactive whiteboard or for independent or paired work at a computer. In addition, the software provides:

- an easy-to-use assessment and allocation system allowing you to track the progress of individual pupils, groups or the whole class
- full planning support to integrate the software into SHM or Heinemann Maths planning
- a teacher's guide with supporting notes for each activity.

It is designed to provide opportunities for pupils to practise and consolidate key mathematical skills following whole class teaching. Alternatively, you can use the software in a whole-class context to encourage collaborative learning.

Other software designed for the interactive whiteboard is in development at the time of publication. See **www.heinemann.co.uk/shm** for announcements about new products.

Using SHM to deliver the Curriculum for Excellence

Learning and teaching

The approach to learning and teaching in **SHM 5**, **SHM 6** and **SHM 7** is complementary to that described in Curriculum for Excellence and is based on the following key ideas:

- daily lessons to build numerical and mathematical understanding
- direct, interactive teaching that encourages pupil involvement
- systematic development of mental calculation strategies
- development of thinking skills through investigative activities and problem solving
- fostering of positive attitudes through the provision of opportunities to discuss, explain, investigate, share ideas, etc. in collaboration with others
- that mathematics should be seen as useful and relevant by making links, where appropriate, with other areas of the curriculum and with children's everyday lives.

Given this approach, the *Teaching Files* are a crucial component of **Scottish Heinemann Maths**, as they provides guidance on:

- the development of each mathematical topic in a clear, systematic way
- teaching, pupil activities and follow-up work to promote deep understanding and develop skills
- the effective use of resources to enhance learning and promote active methodologies
- the use of mental and oral activities to develop and practise mental strategies and techniques.

Direct teaching is essential. It cannot be replaced by use of *Pupil Sheets, Textbooks* and other course materials. The function of such materials is to:

- allow the teacher and the children to check understanding of what has been taught
- provide a record of work completed
- set new challenges where the children can apply the mathematics they have learned in other contexts.

The Scottish Heinemann Maths pupil materials are not designed as a substitute for teaching, with children working through them on their own without prior teaching and discussion. They are designed to enhance learning by supporting teachers, whose role in developing new concepts through direct teaching and interaction with children remains crucial.

Direct teaching and interaction

High quality teaching and interaction are at the heart of **Scottish Heinemann Maths**. This two-way process encourages both teachers and children to be actively engaged in the learning process.

Children are expected to:

- be actively involved in their learning with opportunities for choice and autonomy
- contribute with confidence to *Teaching* activity and *Follow-up* discussions
- be able to explain and justify their chosen strategies and methods and demonstrate understanding of their learning to others.

The *Teaching Files* make suggestions that enable teachers to provide flexible and effective teaching approaches through an appropriate balance of the following methods.

Demonstration – showing and illustrating mathematics using appropriate resource and visual materials.

Instruction – giving information clearly and precisely.

Direction – sharing teaching objectives with the children and making sure they know what they should be learning.

Explanation and illustration – giving accurate, well-paced explanations.

Questioning – using effective questioning techniques to:

- ensure that children are actively involved in their learning
- encourage the children to explain what they are doing
- help the children to consider other possible strategies and methods
- focus the children's attention on new aspects of their learning.

A wide range of open and closed questions is suggested within the *Teaching* activities in **Scottish Heinemann Maths**.

Consolidating – maximising opportunities to reinforce and develop what has been taught through the use of:

- *Starters and other mental activities* to consolidate previous learning
- well-focused *Pupil Activities* and *Pupil Sheets*
- practice and applications in the *Textbooks*
- extension activities in the *Extension Textbooks.*

Discussion and evaluation of children's responses – identifying mistakes and misunderstandings by using:

- oral *Follow-up* activities from the *Teaching Files*
- *Check-ups* and *Topic Assessment* in the *Assessment* books.

This process can help to identify appropriate further teaching.

Summarising – reviewing with the children what has been taught and what the children have learned using *Follow-up* activities.

Differentiation

Scottish Heinemann Maths has been designed to be used with the whole class, groups or individual children. The *Teaching* activities in the *Teaching Files* have been written for group or whole class teaching, but can be adapted for use with individuals.

The following features of **Scottish Heinemann Maths** can help the teacher to plan differentiated programmes:

- the ability to select from the suggested *Teaching* activities and *Pupil Activities* for each section in the *Teaching Files* allows the teacher to exercise professional judgement in planning appropriate programmes to match the children's needs
- notes on *Further Teaching* can suggest alternative approaches for particular children who may require additional reinforcement
- *Pupil Sheets* provide further practice for those who need it. Some of them provide templates for teachers to customise for use with groups or individuals
- the activities within the *Extension Textbooks* provide both lateral and vertical extension. They provide opportunities to extend children's mathematical thinking and problem solving skills.

The teacher should omit pages, parts of pages or questions in the *Textbooks* which are considered inappropriate for specific children. However, all children should have experience of using and applying the mathematics they are learning.

Planning

Starting points

Planning for effective learning involves thinking about:

- the Numeracy and Mathematics Outcomes from Curriculum for Excellence.
- The charts on pages 40–83 summarises the curriculum coverage provided by **SHM 5**, **SHM 6** and **SHM 7**.
- the children's previous experience in mathematics.

This can be found by consulting:

- the children's records of achievement
- the mathematical development charts on pages 96–99.
  - the development of the children's knowledge, skills and understanding.

Information relating to the work contained in **SHM 5**, **SHM 6** and **SHM 7** is found in:

- the pupil record grids on pages 110–116 of this guide
- the summary at the beginning of each new topic in the *Teaching Files*. It describes *Previous work* and gives an *Overview* and a detailed *Development* of the topic. The *Contents* table lists the sections within the topic and all the **Scottish Heinemann Maths** materials associated with them. The *Language* list gives a clear indication of the vocabulary used.

While there are many ways to plan a mathematics programme, and schools will have their own planning formats, **Scottish Heinemann Maths** teaching resources provide examples of a development planner, a block planner and a weekly planner. These are designed to provide a route through the materials.

Development planner

The development planner illustrates a progression in the teaching of each maths topic in **SHM 5**, **SHM 6** and **SHM 7**. The content of **SHM 5**, **SHM 6** and **SHM 7** has been divided into a number of planning units. Detailed planners are included on pages 40–83 of this guide.

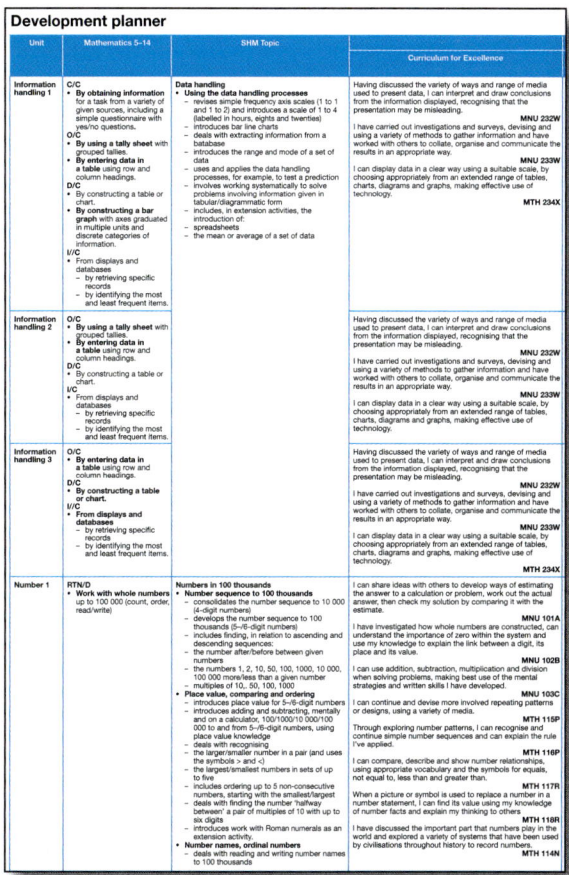

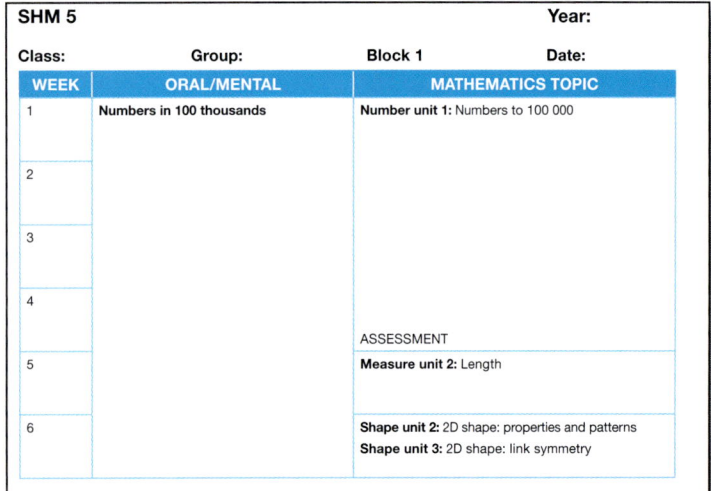

Block planner

The block planner identifies one possible way of organising the content of **SHM 5**, **SHM 6** or **SHM 7** over the course of the year. The year is divided into six 6–week teaching blocks. Detailed planners are included on pages 84–95 of this guide.

Weekly planner

It is often necessary to plan in more detail on a weekly or daily basis. Plans of this type provide an indication of what the teacher hopes to cover during the course of the week. However, given that the mathematics must necessarily build on the needs of the children, an element of flexibility should be built in. This allows for modification to the plan as it is implemented.

The *Schematic* diagrams at the beginning of each section in the *Teaching Files* provide a helpful starting point for this process.

SHM 5 *Teaching File,* page 248
Decimals

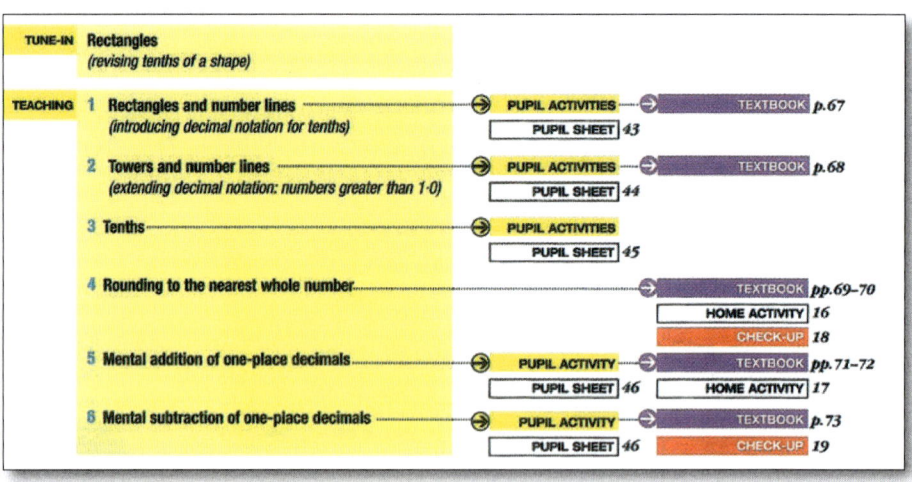

SHM 5 *Teaching File,* page 248
Decimals

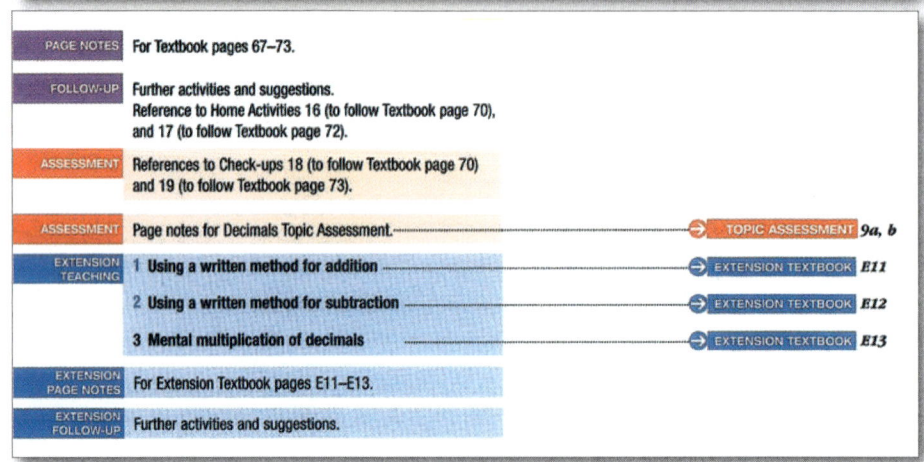

The weekly planner on page 21 shows how *Multiplication* and *Division* might be developed over the course of the week for a Primary 5 class.

Weekly Planner: Multiplication and Division   SHM 5

SECTION	Multiplication by 10, 100			Division by 2–10, 100, 1000	
	Monday	**Tuesday**	**Wednesday**	**Thursday**	**Friday**
TUNE-IN or STARTER	**Tune-in** **Multiplying by 10** (revising multiplication of a two-/three-/four-digit number by 10)	Activity from *Starters and other mental activities* related to multiplying a two-/three-digit number by 10 or a multiple of 10	**Tune-in** **Dividing by 2, 3, 4, 5 and 10** (revising mental division)	Activity from *Starters and other mental activities* related to dividing by 10	Activity from *Starters and other mental activities* related to dividing by 10
TEACHING	**1 Multiplying a two-/three-digit number by 100** *T/F.144–145*	**2 Multiplying a two-digit multiple of 10 by a three-digit multiple of 10** *T.F. 146–147*	**1 Dividing by 6, 7, 8 and 9** *T.F. 177* • Revising mental division	**2 Dividing by 10, 100 and 1000** *T.F. 178–179* • Mental division of three- and four-digit numbers	**3 Divisibility** *T.F. 179–180*
PUPIL ACTIVITIES	**Pupil activities:** 1 Multiplication game 2 Mental multiplication game **Textbook page 36**	**Pupil activity:** 1 Card game **Textbook page 37** Check-up 9	**Pupil activities:** 1 Division dominoes **Pupil Sheet 25** 2 **Pupil Sheet 26** **Textbook pages 48–49**	**Textbook page 50**	**Textbook page 51**
FOLLOW-UP	For **Textbook page 36**	For **Textbook page 37**	For **Textbook pages 48–49**	For **Textbook page 50**	For **Textbook page 51**
REVIEW	**This would include:** – *key areas of concern to build into the next day's teaching* – *any organisation issues to be addressed* – *any child needing particular help.*				

- In the summary at the beginning of each new topic in the *Teaching Files* there is a list of *Resources* required for the topic.

Resources

General
number cards showing three-/four-digit numbers, the 51 to 100 part of a 100-square…

Resource Sheets
27, 51, 52, 53

These are listed under two headings – *General* and *Resource Sheets*.

The *General* resources list the types of practical materials that are normally to be found in Primary 5, 6 or 7 classrooms. These resources should be easily accessible to the children.

The photocopiable *Resource Sheets* that teachers may wish to use in this section of teaching are indicated. These materials are designed to be used with/by the children.

Other resources, such as the necessary *Pupil Sheets, Check-ups, Topic Assessments, Extension* and *Home Activities*, are listed in the *Contents* table for the topic and the schematic diagrams for each section.

Contents

	TEACHING FILE	TEXTBOOK	PUPIL SHEETS	HOME ACTIVITIES	ASSESSMENT	EXTENSION TEXTBOOK
Doubles and near doubles	70–75	10–12		3	CHECK-UP 5	
Addition involving three-digit numbers	76–82	13–17	9	4	CHECK-UP 6	
Addition involving four-digit numbers	83–87	18–21				E2
					TOPIC	

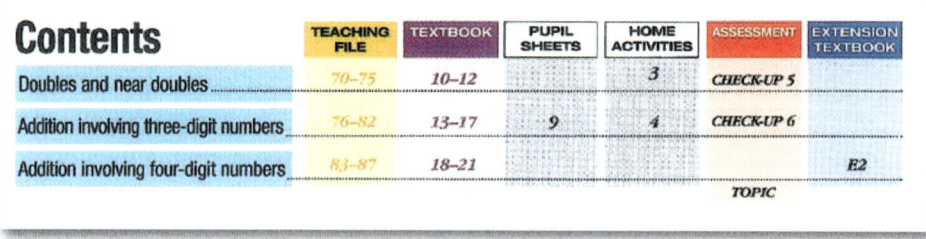

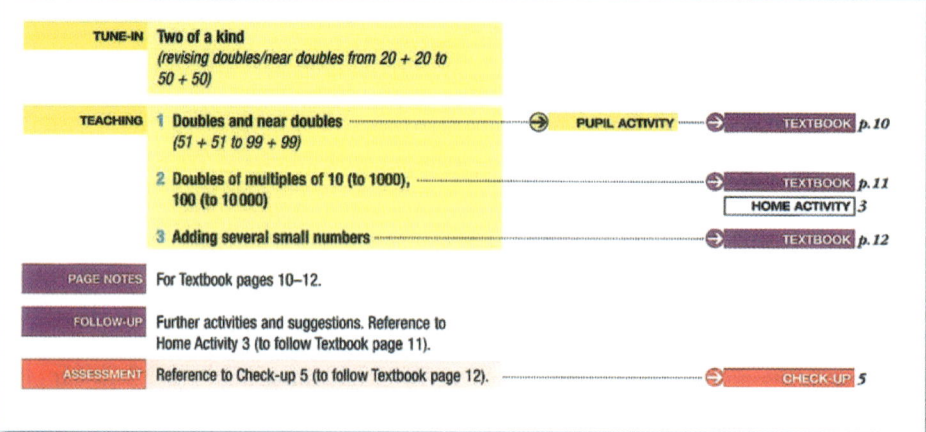

SHM 5 *Teaching File*, page 70 *Addition*

- The materials in **Scottish Heinemann Maths** allow the teacher to structure each mathematics lesson as follows:

  – **Tune-in** (oral and mental work) using the appropriate *Tune-in*, or an activity from the *Starters and other mental activities* section of the *Teaching Files*

- **Teach/Try** (teaching and practice activities) using *Teaching* and *Pupil Activities*, *Pupil Sheets* and appropriate pages of the *Textbooks*, or *Extension Textbooks*, described and illustrated in the *Teaching Files*
- **Talkabout** (a follow-up discussion) using further activities and suggestions, for example, from the *Follow-up* to *Textbook*, or *Extension Textbook* pages as suggested in the *Teaching Files*. The *Follow-up* section also indicates appropriate *Home Activities*.

The teaching suggestions given in the *Teaching Files* do not indicate approximate timings for each part of the lesson. Teachers will use their professional judgement to determine the most appropriate pacing, organisation and specific activities that best meet the needs of the children and the topic.

- The *Tune-in*, *Teaching* and *Pupil Activities* should be completed before the children attempt related *Textbook* or *Extension Textbook* pages.

  When the children are ready to attempt the *Textbook* page it may be necessary for the teacher to discuss some of the following:

  - what the children have to do, focusing on any concerns with language or interpretation of instructions
  - where to find any resource materials they may need
  - how they should set out work or record answers
  - which questions they should complete or omit.

  Teachers, on occasion, may want to ask the children to:

  - do the examples on a *Textbook* page orally without keeping a written record
  - interpret an example in their own words
  - work collaboratively in pairs or small groups with only one child recording the answers.

  Sometimes it may be appropriate to tackle a page as a class or group discussion with each child writing answers as they go along.

- The follow-up discussion is an important part of the lesson. During this time the teacher may wish to:

  - ask children to show and explain their work to other children
  - through discussion, draw together what has been learned and, where appropriate, extend the work
  - provide tasks for the children to complete at home to consolidate their work in class using, for example, a *Home Activity*
  - make connections to other areas of the curriculum or to 'real life' situations.

Delivering the Curriculum for Excellence

Assessment is for Learning

The recommended approaches to assessment and the specific materials provided for this purpose in **Scottish Heinemann Maths** allow teachers to address the following aspects identified in *Assessment is for Learning*:

- Assessment FOR learning – learners are at the centre of the process and are aware of the purpose of their learning, where they are in relation to set goals and have been given clear feedback as to how to move forward
- Assessment AS learning – learners are aware of their own learning needs and work with teachers and peers to assess their strengths and areas for development
- Assessment OF learning – reliable, robust information is gathered in a variety of different ways that are appropriate to the context, the learner and the purpose of the data.

Formative

Much of the assessment of children's learning of mathematics in the primary classroom is of an informal nature and happens on a daily basis. This can be done during many of the activities that children are involved in such as:

- the *Tune-in* or *Starters* and other mental activities
- the main teaching activity
- the *Pupil Activities*, including games, practical activities, *Pupil Sheets* or *Textbooks*
- the *Follow-up* discussion highlighting the main teaching ideas.

These provide formative evidence that the teacher can use to determine the extent of a child's understanding of a particular mathematical concept and inform the nature of any subsequent teaching action. This evidence can be gathered in a number of different ways, for example, by:

- listening to and talking with the children (posing questions and noting responses)
- observing the children (noting individual strengths and needs)
- reviewing the children's written work
- taking account of the results of children's self- or peer-assessment.

However, more focused methods of assessment are also necessary.

Focused assessment

Scottish Heinemann Maths provides:

- *Check-ups* to assess the children's understanding of a section of work they have recently completed
- *Topic Assessments* to assess a mathematical topic
- *Round-ups* to assess the children's understanding of the broad range of concepts and skills they have learned during the course of a year.

It is also possible to use *Textbook* pages to assess concepts taught in the course of interactive lessons. This can be done by reviewing the children's completed work (possibly through discussion) or by observation while the work is in progress.

Check-ups

When the teacher wishes to use a more objective, specific task to check on the children's understanding of a particular section of teaching in mathematics, for example 'Dividing by 8', one of the *Check-ups*, can be used. These are provided in the *Assessment* books and in photocopiable format.

While a *Check-up* will normally be used immediately after a section of work has been completed, a teacher may decide to assess that section in another way. The *Check-up* can then be used at a slightly later stage to determine if knowledge and understanding has been retained. The *Schematics* and *Contents* tables in the summary pages indicate which *Check-up* relates to a particular section of work and when it could be used.

ASSESSMENT	Reference to Check-up 9 (to follow Textbook page 37).		CHECK-UP 9

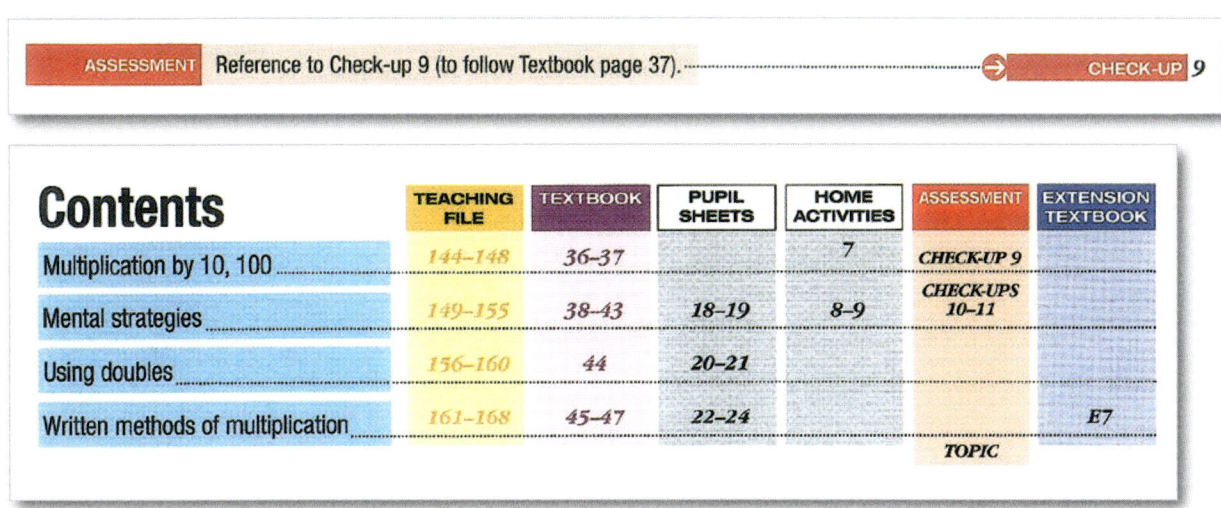

Contents	TEACHING FILE	TEXTBOOK	PUPIL SHEETS	HOME ACTIVITIES	ASSESSMENT	EXTENSION TEXTBOOK
Multiplication by 10, 100	144–148	36–37		7	CHECK-UP 9	
Mental strategies	149–155	38–43	18–19	8–9	CHECK-UPS 10–11	
Using doubles	156–160	44	20–21			
Written methods of multiplication	161–168	45–47	22–24			E7
					TOPIC	

The teaching notes give details of the mathematics covered and the relevant *Textbook* pages for each *Check-up*.

ASSESSMENT CHECK-UP 9	Check-up 9 can be used to assess the work on *Multiplying by 100/a multiple of 100*. It is related to Textbook pages 36–37.

These notes follow the *Page Notes* and suggestions for *Follow-up* activities.

The *Check-ups* provide a valuable record of achievement/attainment that can:

- highlight where further teaching or consolidation may be necessary
- be shared and discussed with children so that they are aware of their progress
- provide a basis for self-assessment
- be used as a focus of discussion with parents
- be transferred along with other evidence to another teacher or school.

Topic Assessments

The **SHM 5** *Assessment* book contains 9 *Topic Assessments*, the **SHM 6** *Assessment* book contains 15 *Topic Assessments* and the **SHM 7** *Assessment* book contains 17 *Topic Assessments*. These assess the work related to a specific number or time topic, such as Multiplication:

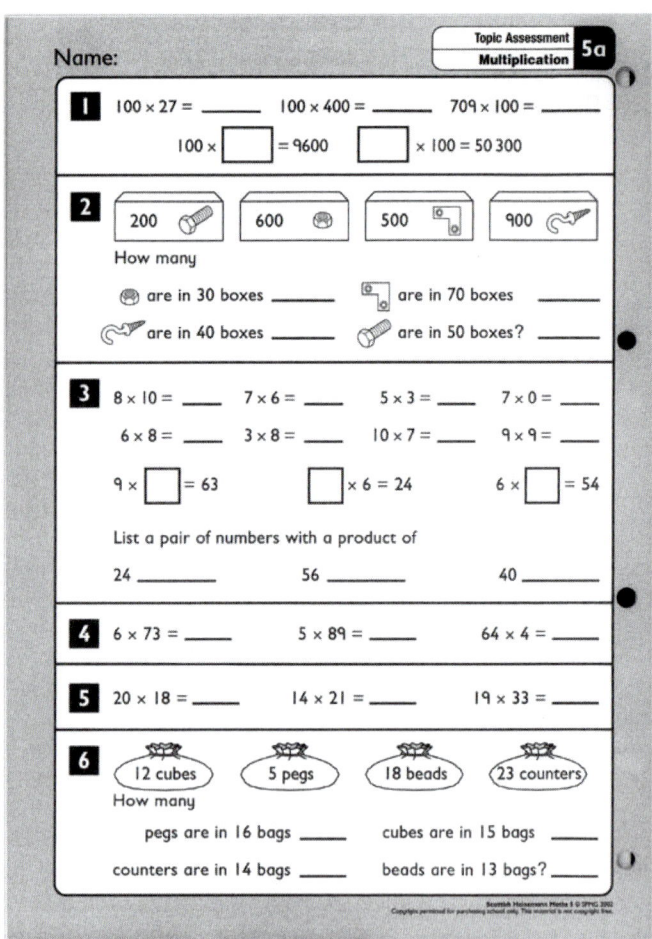

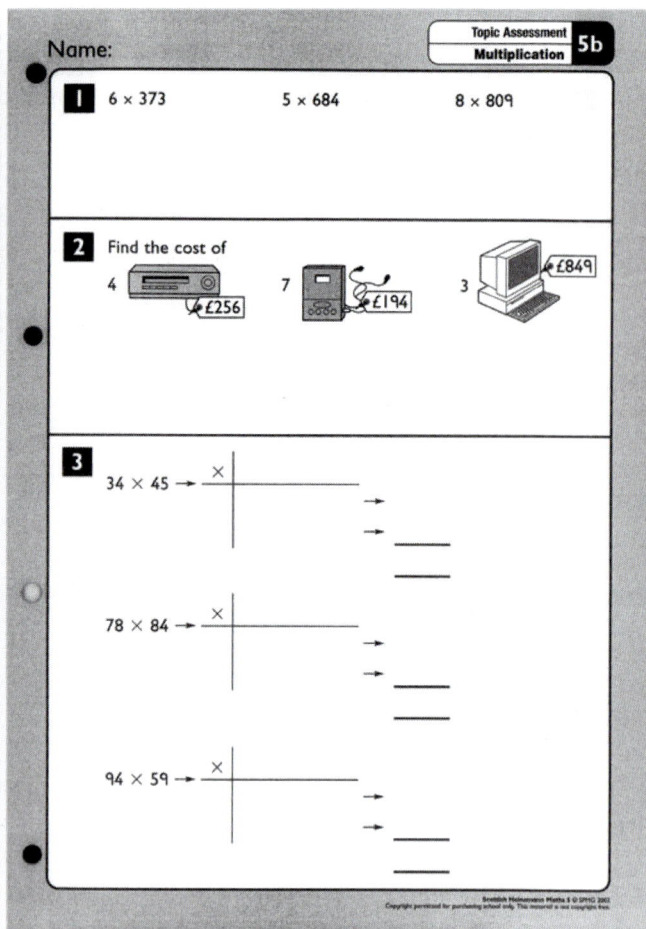

These pages are easily identified by the solid coloured border around the whole page. This type of assessment is normally used when a whole topic has been completed.

The *Schematic* diagram at the start of each section indicates where these assessments occur and when they could be used.

The teaching notes give details of:

– the mathematics covered and the relevant *Textbook* pages

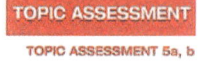

TOPIC ASSESSMENT

TOPIC ASSESSMENT 5a, b

Topic Assessment 5a, b assesses a range of work on *Multiplication* associated with Textbook pages 36–47.

– specific equipment or materials required
– what each question is assessing and any common errors that are likely to occur
– brief suggestions about how to deal with repeated errors
– references to the appropriate section of the *Teaching File* to return to if further teaching or consolidation is deemed necessary.

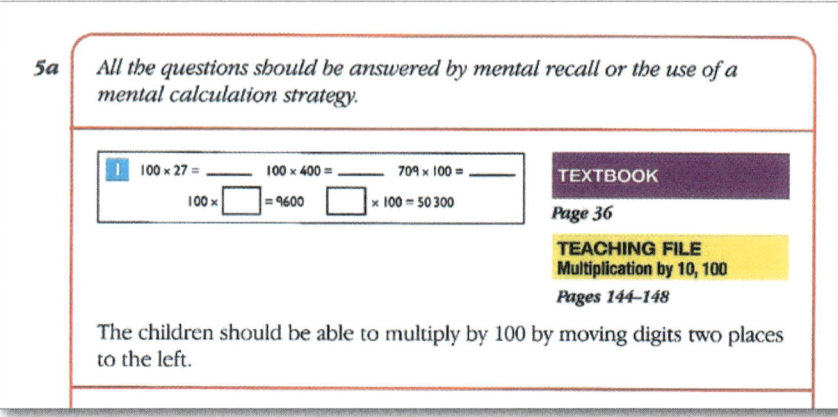

Round-up assessments

The *Round-up* assessments check on the full range of mathematical topics. These *Round-ups* are provided in the *Assessment* books and are also available in photocopiable format.

SHM 5

- *Round-up 1* provides assessment at Level C for each of the mathematical areas of Number, Measure, Shape and Data handling and can be used at appropriate times during the year.

- *Round up 2* and *Round-up 3* are end-of-year assessments that contain questions covering a range of 'mixed' mathematic:

  - *Round-up 2* includes questions related to Numbers in 100 thousands, Addition and Subtraction to 1000, Multiplication, Division, Fractions, Decimals, Time, Length, Capacity, 3D shape, Position, movement and angle, and Data handling.

  - *Round up 3* includes questions related to Numbers in 100 thousands, Addition and Subtraction beyond 1000, Multiplication, Division, Fractions, Decimals, Time, Length and Weight.

SHM 6 and **SHM 7**

- The end of year *Round-up* assessment checks on the full range of the mathematics covered.

Completion of the *Round-ups* provides summative evidence of how well children are progressing in relation to the Curriculum for Excellence Numeracy and Mathematics Outcomes appropriate to their stage of development.

Using the assessment materials

The *Check-ups, Topic Assessments* and *Round-ups* provide useful information on:

- an individual child's progress, noting, in particular, areas of success but also highlighting any specific areas of difficulty

- how the class or groups of children are progressing, indicating success or common difficulties that have emerged which may require attention.

For example, discussion with the children about their work on a particular *Check-up* may help to establish why specific questions proved more difficult. From such discussions important teaching points may emerge and children can be encouraged to view any difficulties or misunderstandings, not as judgements, but as part of the learning process.

One copy of the *Check-ups* could be used to record specific comments about common difficulties. For example, using a highlighter pen to indicate questions where a significant number of children experienced difficulty, or by circling questions where no errors occurred.

Delivering the Curriculum for Excellence

Assessment record grids

Assessment record grids are provided on pages 100–102 of this guide to help record class, group or individual coverage of the *Check-ups*, *Topic Assessments* and *Round-ups* completed. Ways of recording coverage might include a tick (✓) or a qualitative indicator of how well a specific assessment was completed. This could be done by shading part or all of an individual box. For example:

no errors made

few errors made

a significant number of errors made.

Level C, D and E class record grids

Pages 103–109 provide a simple checklist of all the attained targets for Level C, D and E. Space is provided to note work which has been attempted and the quality of the performances of the individuals within the class. It enables the teacher to monitor the progress of the whole class.

Record of work grids

Record of work grids are provided on pages 110–116 of this guide. These grids can be used to show:

- when work has been completed
- how well the work has been completed, for example by using a code as above.

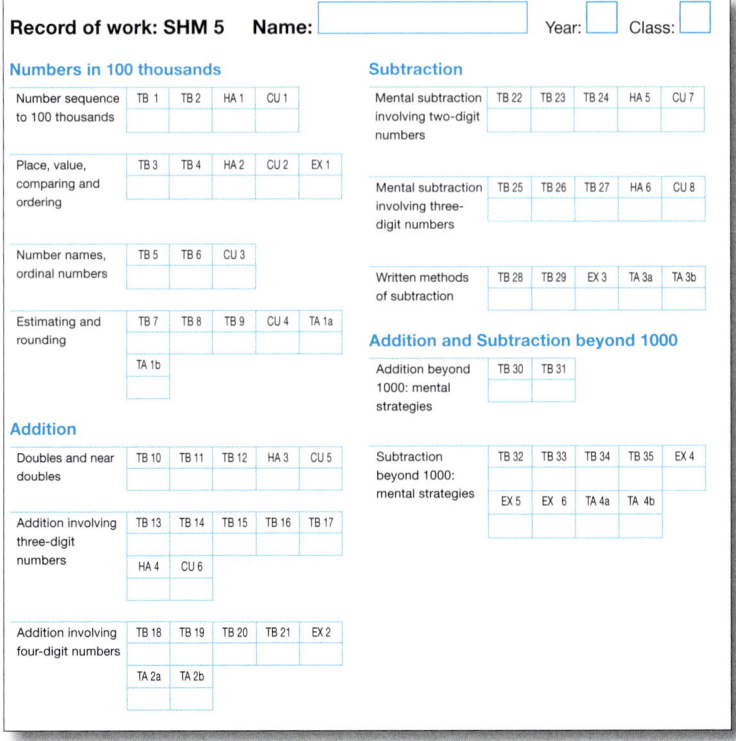

Using your Planning and Teaching CD-ROM

The CD-ROM accompanying this book contains all of the charts and tables within the book as editable Word documents, so you can amend and adapt them as you wish. It also contains a number of video clips and photographs, so that you can see examples of A Curriculum for Excellence being delivered by a school using **Scottish Heinemann Maths**. For your convenience, the CD–ROM also contains the Programmes of Study.

Videos on the CD-ROM

There are several short clips on the CD-ROM, filmed at Jordanhill Primary School, 45 Chamberlain Road, Jordanhill, Glasgow, G13 1SP in late 2007 and early 2008. The publisher and authors wish to thank Gordon Smith, Betty Edmonstone, Liz Edmonstone, Emma Bryson, Gillian Kennedy, Lynne Doyle and all the pupils who took part for their help in compiling this short introduction to how a school using **SHM** has begun to implement A Curriculum for Excellence. You could use the clips as a starting point for ideas about delivering A Curriculum for Excellence in your own school, and also as part of a presentation during a staff meeting.

List of video clips on the CD-ROM

- Children enjoying maths

    Emma Bryson, Primary 5 teacher, on how her children engage and enjoy maths lessons.
- The challenges of A Curriculum for Excellence

    Gordon Smith, Head Teacher at Jordanhill Primary school, talks about the greatest challenges presented by A Curriculum for Excellence.
- Integrating problem-solving into maths

    Emma Bryson talks about how she integrates problem-solving into her maths lessons.
- Effective questioning and assessment techniques in maths

    Emma discusses Assessment is for Learning within her maths lessons.
- Supporting lower attainers

    How children performing at a lower level are supported in Primary 5 at Jordanhill.
- Teaching a lesson using **Scottish Heinemann Maths**

    *Gillian Kennedy, Primary 5 teacher, talks about how she has used **SHM** during a recent lesson.*
- Assessment is for Learning, differentiation and ICT in a **Scottish Heinemann Maths** classroom

    Gillian Kennedy talks about her next maths lesson.
- Assessment is for Learning strategies within a maths lesson

    Gillian Kennedy demonstrates AifL strategies within one of her maths lessons.
- Active learning within a classroom environment

    *A Primary 5 class carry out a **SHM** activity in their classroom.*
- Active learning outside of the classroom environment

    *A Primary 5 class carry out a **SHM** activity outside of their classroom.*

- Cross-curricular clip

    A group of Primary 5 children transfer some of their mathematical skills to an art lesson.

- Using **SHM** pupil software

    *A Primary 5 class use **SHM** pupil software within a maths lesson.*

- Peer assessment

    A P5 class assess their own understanding of a recent maths topic.

Photographs on the CD-ROM

The photographs may be useful for illustrating, for your own information or for colleagues, how different resources, including **Scottish Heinemann Maths**, can be used as part of the implementation of A Curriculum for Excellence.

List of photographs on the CD-ROM

1. How **SHM** workbooks can help to provide a permanent record of children's recording

2. A creative use of **SHM** lesson ideas and resources to make a classroom display

3. An example of active learning within **SHM** – a model children have made during a **SHM** lesson

4. Some of the strategies used at Jordanhill to share pupils' targets and successes

5. More active learning based on **SHM** materials

6. Children using peer-to-peer coaching and assessment

7. The **SHM** pupil software in action

8. How **SHM** interactive whiteboard resources can be used to enhance maths teaching and learning.

Problem Solving Activities

Introduction

This section contains four problem solving activities which are intended to be dipped into from time to time. They make a contribution to the development of problem solving strategies and provide opportunities for children to use and apply the mathematical facts and techniques they have learned in a variety of contexts, and supplement problem solving in the workbooks of **Scottish Heinemann Maths**.

The activities are presented as notes for teachers and photocopiable sheets for children. The photocopiable sheets are of two types: some are expendable *worksheets* on which children record their findings; others are presented as problem solving *activities* on pages which do not have a fill-in format.

The notes for teachers provide:

- information about the curriculum coverage and the materials required
- advice about using the activities, possible strategies, solutions and additional activities.

Using the problem solving activities

- Each activity can be used at any time after the children are familiar with the underlying mathematical ideas. This is indicated in the teaching notes for the activity.
- Although it is possible for the activities to be attempted on an individual basis, there are clear benefits to be derived from having the children work collaboratively in pairs or groups of three.
- In starting each activity, the teacher's role is to set the scene and discuss the activity with the children without telling them how to solve the problem, making sure they are clear about its requirements. As far as possible, the children should be encouraged to choose a strategy for themselves and carry out the task in their own way.

Assessment

- An important purpose of assessment is to reveal strengths and weaknesses which individuals have when involved in problem solving processes. You should consider, where appropriate, if the child:
  - contributes to the initial discussion of the problem
  - can identify key words and phrases
  - can restate the problem in his or her own words
  - is able to choose the materials to use
  - shows determination to overcome any difficulties which may arise
  - suggests and uses a suitable strategy
  - can report orally the conclusions
  - can record findings in written or pictorial form
  - works collaboratively.

Some of the children's behaviours and conclusions may provide a focus for follow-up work.

Find the price

MATERIALS

A calculator could be useful in question 2.

IMPLEMENTATION

- In question 1, the children should look for items and prices

    — common to all three sets – for example, and

    — common to two sets – for example, and

    — which appear in only one set – for example
    and
    apple pie,

- Using a process of elimination, the children should find the price of each item as shown in the table.

Item	Price
crisps	24p
orange	32p
ice-cream	50p
gums	17p
burger	75p
apple pie	68p
chips	92p

- In question 2, the children should realize that it is unlikely that the chips, apple pie or burger will be included because of their prices.

    Trial and improvement methods, perhaps using a calculator, should lead to the solution:

gums		crisps		orange		ice-cream	
17p	+	24p	+	32p	+	50p	= £1·23

    Possible strategies: trial and improvement; reason logically; be systematic

Find the price

Name _____

1 Find the price of each item. Write your answers in the table.

Item	Price
crisps	

2 Four of the items cost a total of £1·23. What are they?

_____   _____

_____   _____

Calculator capers

MATERIALS

Calculator

IMPLEMENTATION

- The children should realize, in question 1, that they should arrange the digits
    - in decreasing order for the largest number – for example, 8754 for the square illustrated
    - in increasing order for the smallest number – for example, 4578.

    The difference between the largest and the smallest numbers is always **4176**, as shown:

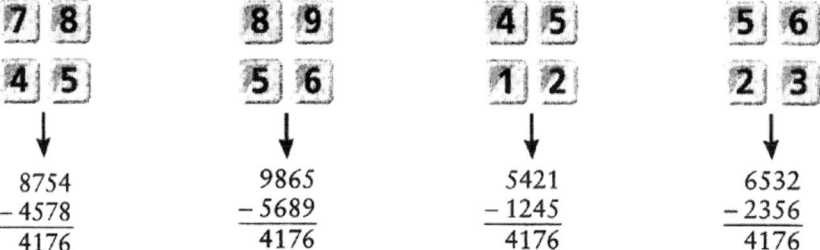

- In question 2(b), the children should notice that the difference between the answers is always 3. Here are the products for the possible squares:

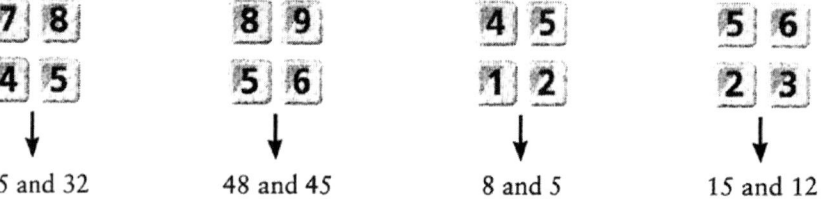

- The children may suggest a variety of patterns. These might include
    - the difference between the top number and the number below it, in each square, is always 3
    - the difference between the totals of the rows is always 6.

    Also, if two-digit numbers are formed by the digits
    - in each row, then the difference between the numbers is always 33
    - in opposite corners, then the difference between these numbers is always 9.

Possible strategies: be systematic; use a diagram; look for a pattern

Calculator capers

1 Number squares

(a) Use this square of 4 number keys.

- ■ Make the largest 4-digit number.
- ■ Make the smallest 4-digit number.
- ■ Find the difference.

(b) Do this again for other squares of 4 number keys. What do you notice?

2 Corners

(a) Use this square again.

Multiply the numbers in the opposite corners together. Write both answers.

 and

(b) Do this again for other squares. What do you notice?

3 Investigate your number squares for other patterns. Write about what you find.

Delivering the Curriculum for Excellence

Threes and fours

MATERIALS

None

IMPLEMENTATION

- In question 1 the children have to draw different shapes with an area of 4 cm^2. They are likely to tackle this by drawing more obvious shapes involving whole squares first.
  For example,

  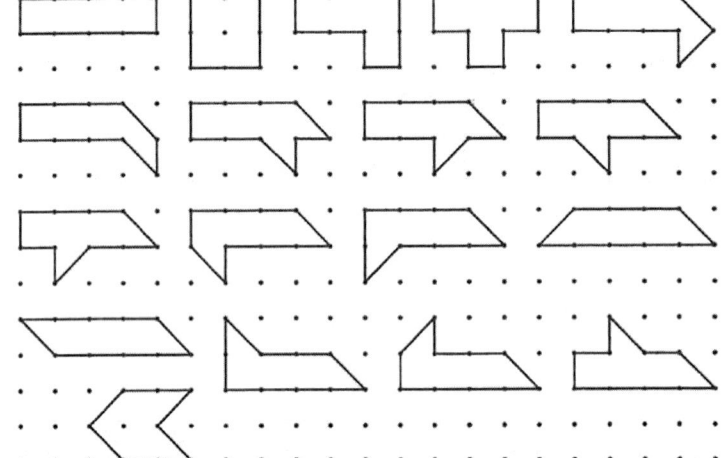

  They should then use half squares: for example, making shapes with 3 whole squares and 2 half squares and shapes with 2 whole squares and 4 half squares.
- Some children may give shapes which they think to be different, but which are in fact rotations or reflections of shapes that they have already produced.
- Here are some shapes with an area of four squares.

- Here are some shapes with an area of three squares.

Possible strategies: draw a diagram; trial and improvement

Threes and fours

Name _____

1 The area of each of these shapes is 4 square centimetres. Join dots to make 10 different shapes each with an area of 4 cm².

2 Draw 10 different shapes each with an area of 3 cm².

Direct routes

MATERIALS

Coloured pencils

IMPLEMENTATION

- The children should draw routes from and to the centres of the planets. Although it would be helpful, it is not necessary to draw straight lines.
- In question 1(b), the children should assume that each route they draw goes 'both ways'. For example, having drawn a route from Anno to Cosmo, they should not draw another one from Cosmo to Anno. By using a particular colour for drawing all the new routes from Cosmo, then another colour for all the new routes from Zolga and so on, the children will find it easier to count the number of routes between the five planets: that is, 4 + 3 + 2 + 1 = 10 routes.
- In question 2, the children draw and name a planet of their own. Although this planet could be placed anywhere on the picture, its best location would be in the space at the bottom of the picture. This would allow the new routes they draw to be clearly seen. Six planets give a total of 5 + 4 + 3 + 2 + 1 = 15 routes.

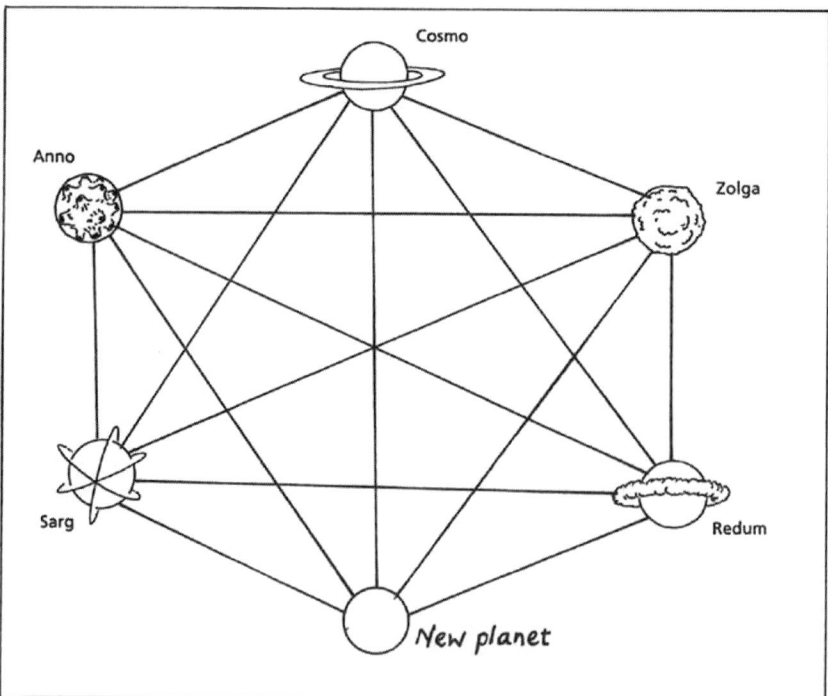

Possible strategies: draw a diagram; work systematically; look for a pattern

Direct routes

Name _____

You need coloured pencils.

Cosmo

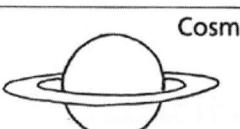

Anno

Zolga

Sarg

Redum

1 (a) Draw routes from Anno to each of the other planets.
Use a coloured pencil.

(b) Repeat for each of the other planets. Use a different colour each time.
Do not repeat routes already shown.

(c) How many routes are there altogether
between the five planets? _____ routes

2 Draw another planet. Find the
number of routes between all six planets. _____ routes

Development planner

Unit	Mathematics 5–14	SHM Topic	Curriculum for Excellence
Information handling 1	**C/C** • **By obtaining information** for a task from a variety of given sources, including a simple questionnaire with yes/no questions. **O/C** • **By using a tally sheet** with grouped tallies. • **By entering data in a table** using row and column headings. **D/C** • **By constructing a table or chart**. • **By constructing a bar graph** with axes graduated in multiple units and discrete categories of information. **I/C** • **From displays and databases** – by retrieving specific records – by identifying the most and least frequent items.	**Data handling** • **Using the data handling processes** – revises simple frequency axis scales (1 to 1 and 1 to 2) and introduces a scale of 1 to 4 (labelled in hours, eights and twenties) – introduces bar line charts – deals with extracting information from a database – introduces the range and mode of a set of data – uses and applies the data handling processes, for example, to test a prediction – involves working systematically to solve problems involving information given in tabular/diagrammatic form – includes, in extension activities, the introduction of: – spreadsheets – the mean or average of a set of data.	Having discussed the variety of ways and range of media used to present data, I can interpret and draw conclusions from the information displayed, recognising that the presentation may be misleading. **MNU 232W** I have carried out investigations and surveys, devising and using a variety of methods to gather information and have worked with others to collate, organise and communicate the results in an appropriate way. **MNU 233W** I can display data in a clear way using a suitable scale, by choosing appropriately from an extended range of tables, charts, diagrams and graphs, making effective use of technology. **MTH 234X**
Information handling 2	**O/C** • **By using a tally sheet** with grouped tallies. • **By entering data in a table** using row and column headings. **D/C** • **By constructing a table or chart**. **I/C** • **From displays and databases** – by retrieving specific records – by identifying the most and least frequent items.		Having discussed the variety of ways and range of media used to present data, I can interpret and draw conclusions from the information displayed, recognising that the presentation may be misleading. **MNU 232W** I have carried out investigations and surveys, devising and using a variety of methods to gather information and have worked with others to collate, organise and communicate the results in an appropriate way. **MNU 233W** I can display data in a clear way using a suitable scale, by choosing appropriately from an extended range of tables, charts, diagrams and graphs, making effective use of technology. **MTU 234X**
Information handling 3	**O/C** • **By entering data in a table** using row and column headings. **D/C** • **By constructing a table or chart**. **I/C** • **From displays and databases** – by retrieving specific records – by identifying the most and least frequent items.		Having discussed the variety of ways and range of media used to present data, I can interpret and draw conclusions from the information displayed, recognising that the presentation may be misleading. **MNU 232W** I have carried out investigations and surveys, devising and using a variety of methods to gather information and have worked with others to collate, organise and communicate the results in an appropriate way. **MNU 233W** I can display data in a clear way using a suitable scale, by choosing appropriately from an extended range of tables, charts, diagrams and graphs, making effective use of technology. **MTH 234X**
Number 1	**RTN/D** • **Work with whole numbers** up to 100 000 (count, order, read/write)	**Numbers in 100 thousands** • **Number sequence to 100 thousands** – consolidates the number sequence to 10 000 (4-digit numbers) – develops the number sequence to 100 thousands (5-/6-digit numbers) – includes finding, in relation to ascending and descending sequences: – the number after/before between given numbers – the numbers 1, 2, 10, 50, 100, 500, 1000, 10 000, 100 000 more/less than a given number – multiples of 10, 50, 100, 1000. • **Place value, comparing and ordering** – introduces place value for 5-/6-digit numbers – introduces adding and subtracting, mentally and on a calculator, 100/1000/10 000/ 100 000 to and from 5-/6-digit numbers, using place value knowledge – deals with recognising: – the larger/smaller number in a pair (and uses the symbols > and <) – the largest/smallest numbers in sets of up to five – includes ordering up to 5 non-consecutive numbers, starting with the smallest/largest – deals with finding the number 'halfway between' a pair of multiples of 10 with up to six digits – introduces work with Roman numerals as an extension activity. • **Number names, ordinal numbers** – deals with reading and writing number names to 100 thousands	I can share ideas with others to develop ways of estimating the answer to a calculation or problem, work out the actual answer, then check my solution by comparing it with the estimate. **MNU 101A** I have investigated how whole numbers are constructed, can understand the importance of zero within the system and use my knowledge to explain the link between a digit, its place and its value. **MNU 102B** I can use addition, subtraction, multiplication and division when solving problems, making best use of the mental strategies and written skills I have developed. **MNU 103C** I can continue and devise more involved repeating patterns or designs, using a variety of media. **MTH 115P** Through exploring number patterns, I can recognise and continue simple number sequences and can explain the rule I've applied. **MTH 116P** I can compare, describe and show number relationships, using appropriate vocabulary and the symbols for equals, not equal to, less than and greater than. **MTH 117R** When a picture or symbol is used to replace a number in a number statement, I can find its value using my knowledge of number facts and explain my thinking to others. **MTH 118R** I have discussed the important part that numbers play in the world and explored a variety of systems that have been used by civilisations throughout history to record numbers. **MTH 114N**

SHM Resources					Assessment		Other Resources	Date	Comment
Teaching File page	Textbook	Extension Textbook	Pupil Sheet	Home Activity	Check-Up	Topic Assessment			
400–409	119–123	E21–E22	74			41			
36–41	1–2		1–5	1	1				
42–52	3–4	E1	6–7	2	2				
53–57	5–6		8		3				

Development planner

Unit	Mathematics 5–14	SHM Topic	Curriculum for Excellence
Number 1 (cont.)	**RN/D** • **Round any number** to the nearest appropriate whole number, 10 or 100. **PS/C** • **Work with patterns and sequences** within and among multiplication tables.	– extends ordinal numbers and their associated notation to include any number to 100, for example, twenty-first (21st), fifty-second (52nd), sixty-third (63rd). • **Estimating and rounding** – revises estimation of a multiple of 10 from its position on a 0–100 number line and extends this to: – a multiple of 100, on a 0–1000 line – a multiple of 5, on a 0–50 line – a multiple of 20, on a 0–200 line – revises rounding a 3-digit number to the nearest 100 and to the nearest 10 – introduces rounding a 4-digit number to the nearest 1000, nearest 100 and nearest 10. **ASSESSMENT** • **Number properties** – explores number patterns with grids, and investigates further odd and even numbers – develops finding rules for number sequences – introduces using a calculator to generate number sequences – includes multiples of 6, 7, 8, 9 and 11 and common multiples – develops working with factors to include factor pairs and prime numbers – uses factors in multiplication and division – introduces as extension work, spare numbers, negative numbers and problem-solving activities.	I have investigated how whole numbers are constructed, can understand the importance of zero within the system and use my knowledge to explain the link between a digit, its place and its value. **MNU 102B** I can share ideas with others to develop ways of estimating the answer to a calculation or problem, work out the actual answer, then check my solution by comparing it with the estimate. **MNU 101A** I can use my knowledge of rounding to routinely estimate the answer to a problem, then after calculating, decide if my answer is reasonable, sharing my solution with others. **MNU 201A** Having explored the patterns and relationships in multiplication and division, I can investigate and identify the multiples and factors of numbers. **MTH 207E** I can continue and devise more involved repeating patterns or designs, using a variety of media. **MTH 115P** Through exploring number patterns, I can recognise and continue simple number sequences and can explain the rule I've applied. **MTH 116P** I can compare, describe and show number relationships, using appropriate vocabulary and the symbols for equals, not equal to, less than and greater than. **MTH 117R** When a picture or symbol is used to replace a number in a number statement, I can find its value using my knowledge of number facts and explain my thinking to others. **MTH 118R** I can show my understanding of how the number line extends to include numbers less than zero and have investigated how these numbers occur and are used. **MNU 206D**
Number 2	**AS/C** • Mentally for one digit to whole numbers up to three digits; beyond in some cases involving multiples of 10. • Without a calculator for whole numbers with two digits added to three digits. • In applications in number, measurement and money to £20. **AS/D** • Mentally for 2-digit whole numbers, beyond in some cases, involving multiples of 10 or 100. • Without a calculator, for 4 digits with at most two decimal places. • With a calculator, for four digits with at most two decimal places. **FE/C** • **Use a simple 'function machine' for operations** involving doubling, halving, adding and subtracting.	**Addition** • **Doubles and near doubles** – extends the work on doubles/near doubles to include: – all numbers from 51 to 99, for example: 51 + 51, 77 + 78 – multiples of 10 to 1000, for example, 640 + 640 – multiples of 100 to 10 000, for example 8200 + 8200 – consolidates and develops strategies for adding several small numbers. **Addition involving three-digit numbers** – consolidates and develops mental addition of a 3-digit number and a 2-digit multiple/near of 10 – introduces mental addition of any 3-digit number and any 2-digit number, for example, 325 + 57, 631 + □ = 700 – introduces mental addition of 3-digit multiples of 10, not bridging 1000, for example: 370 + 140, 120 + 310 +180 – deals with mental addition of a 3-digit multiple of 10 and a 3-digit number, for example: 320 + 247, 123 + □ = 363 – introduces mental addition of 3-digit numbers bridging a multiple of 10, for example, 243 + 216, and then bridging a multiple of 100, for example: 384 + 142 – consolidates a standard written method of addition of 3-digit numbers. **Addition involving 4-digit numbers** – introduces addition of a 4-/5-digit number and a 2-digit multiple of 10, for example: 3627 + 70, 12 605 + 80 – deals with addition of a 3-/4-digit number and a 3-digit multiple of 100, bridging a multiple of 1000, for example: 926 + 400, 2682 + 500 – develops the use of a standard written method to examples involving the addition of 4-digit numbers – provides opportunities to use and apply skills in mental and written addition and the use of a calculator. **ASSESSMENT** **Addition beyond 1000** • **Mental strategies** – deals with adding 3-/4-digit multiples of 100 bridging a multiple of 1000, for example: 800 + 300, 400 + 500 + 700, 3600 + 900 – extends work on addition doubles to finding doubles of multiples of 100 to 5000 + 5000, for example: double 3500, 2700 + 2700 – includes finding what must be added to a four-digit multiple of 100 to make the next multiple of 1000, for example: 3200 + □ = 4000 – deals with adding: – 2-/3-digit numbers to multiples of 1000 and 4-digit multiples of 100 (2000 + 35, 7000 + 261, 3400 + 92, 1600 + 236) – 2-digit multiples of 10 and multiples of 100 to any 4-digit number (3265 + 30, 4371 + 400) – provides opportunities for using and applying the above strategies – includes extension work involving addition of 4-digit numbers with no bridging, for example: 2634 + 1253.	I can use addition, subtraction, multiplication and division when solving problems, making best use of the mental strategies and written skills I have developed. **MNU 103C** Having determined which calculations are needed, I can solve problems involving whole numbers using a range of methods, sharing my approaches and solutions with others. **MNU 203C** Through exploring number patterns, I can recognise and continue simple number sequences and can explain the rule I've applied. **MTH 116P** I can compare, describe and show number relationships, using appropriate vocabulary and the symbols for equals, not equal to, less than and greater than. **MTH 117R** When a picture or symbol is used to replace a number in a number statement, I can find its value using my knowledge of number facts and explain my thinking to others. **MTH 118R**

SHM Resources					Assessment		Other Resources	Date	Comment
Teaching File page	Textbook	Extension Textbook	Pupil Sheet	Home Activity	Check-Up	Topic Assessment			
58–64	7–9			4		1a, b			
272–286	74–76	E14–E17	47–52						
70–75	10–12			3	5				
76–82	13–17		9	4	6				
83–87	18–21	E2				2a, b			
122–129	30–31		14–15						

Delivering the Curriculum for Excellence

Development planner

Unit	Mathematics 5–14	SHM Topic	Curriculum for Excellence	

| Number 3 | **AS/C** | **Subtraction** | I can share ideas with others to develop ways of estimating the answer to a calculation or problem, work out the actual answer, then check my solution by comparing it with the estimate. |
| | | | **MNU 101A** |

AS/C
- Mentally for one digit from whole numbers up to three digits; beyond in some cases involving multiples of 10.
- Mentally for subtraction by 'adding on'.
- Without a calculator for whole numbers with two digits subtracted from three digits.
- In applications in number, measurement and money to £20.

AS/D
- Mentally for 2-digit whole numbers, beyond in some cases, involving multiples of 10 or 100.
- Without a calculator, for four digits with at most two decimal places.
- With a calculator, for four digits with at most two decimal places.

Subtraction
- **Mental subtraction involving two-digit numbers**
  - revises mental subtraction of a 2-digit number from a 2-digit number
  - revises mental subtraction of a 2-digit multiple/near multiple of 10 from a 3-digit number, with no bridging of a multiple of 100 (570 – 40, 685 – 60, 390 – 51. 457 – 38), and then deals with examples which bridge a multiple of 100 (320 – 80, 435 – 60, 226 – 42, 544 – 79)
  - introduces mental subtraction of any 2-digit number from a 3-digit number, initially using examples which only bridge a multiple of 10 (473 – 46), and then with examples which bridge a multiple of 100 (342 – 75).

Mental subtraction involving three-digit numbers
  - revises subtracting a multiple of 100 from a 3-digit number (437 – 200)
  - revises subtracting 3-digit multiple of 10 with no bridging (650 – 220) and introduces subtracting a multiple of 10 from a 3-digit number, with no bridging (742 – 210)
  - introduces subtracting 3-digit multiples of 10 with bridging (320 – 180)
  - revises finding small differences between number on 'either side' of the same multiples of 100 (503 – 495, 810 – 791) and extends this to differences between numbers 'just over' or 'just under' **different** multiples of 100 (904 – 398, 407 – 196)
  - includes using and applying calculator skills to solve problems involving the link between addition and subtraction.

Written methods of subtraction
  - revises informal, expanded and standard written methods of subtraction involving 2-digit numbers
  - extends the use of a standard written method to subtraction involving 3-digit numbers
  - explores a further alternative form of written subtraction involving complimentary addition
  - includes using and applying calculator skills in problem solving
  - provides extension activities using a calculator dealing with:
    - subtraction involving 4-digit numbers
    - mixed addition and subtraction problems.

Subtraction beyond 1000
- **Mental strategies**
  - deals with subtracting a 3-digit multiple of 100 from a 4-digit multiple of 100, bridging a multiple of 1000, for example, 3100 – 800
  - introduces subtracting a single digit from a 4-digit number, bridging a multiple of 10, for example, 1600 – 6, 3216 – 8
  - deals with finding small differences between 4-digit numbers on either side of a multiple of 10, for example, 1372 – 1368, 5006 – 4994
  - uses and applies mental strategies for addition and subtraction of 4-digit numbers
  - includes extension work involving subtraction of 4-digit numbers with no exchange for example, 5759 – 3416.

Curriculum for Excellence (Number 3):

I can share ideas with others to develop ways of estimating the answer to a calculation or problem, work out the actual answer, then check my solution by comparing it with the estimate.
MNU 101A

I have investigated how whole numbers are constructed, can understand the importance of zero within the system and use my knowledge to explain the link between a digit, its place and its value.
MNU 102B

I can use addition, subtraction, multiplication and division when solving problems, making best use of the mental strategies and written skills I have developed.
MNU 103C

I can use my knowledge of rounding to routinely estimate the answer to a problem, then after calculating, decide if my answer is reasonable, sharing my solution with others.
MNU 201A

I have extended the range of whole numbers I can work with and having explored how decimal fractions are constructed, can explain the link between a digit, its place and its value.
MNU 202B

Having determined which calculations are needed, I can solve problems involving whole numbers using a range of methods, sharing my approaches and solutions with others.
MNU 203C

Having explored the patterns and relationships in multiplication and division, I can investigate and identify the multiples and factors of numbers.
MTH 207E

Number 4

MD/C
- Mentally within the confines of all tables to 10.
- Mentally for any 2- or 3-digit whole number by 10.
- Without a calculator for 2-digit whole numbers by any single digit whole number.
- In applications in number, measurement and money to £20.

MD/D
- Mentally for whole numbers by single digits.
- Mentally for 4-digit numbers including decimals by 10 or 100.
- Without a calculator for four digits with at most two decimal places by a single digit.
- With a calculator for four digits with at most two decimal places by a whole number with two digits.
- In applications in number, measure and money

Multiplication
- **Multiplication by 10, 100**
  - revises multiplication of a 2-/3-/4-digit number by 10
  - introduces multiplication of a 2-/3-digit number by 100, for example: 56 × 100 = ☐   32 × ☐ = 3200
  - introduces multiplication of a 2-digit multiple of 10 by a 3-digit multiple of 100, for example: 30 x 400.
- **Mental strategies**
  - consolidates the multiplication tables
  - extends multiplying a 2-digit number by a single digit to include examples which bridge a multiple of 10, for example:
    - 3 × 26 → (3 × 20) + (3 × 6) → 60 + 18 = 78
  - revises multiplication of a 2-digit number by 20 and introduces mental multiplication by 19 and 21
  - includes simple problems using language associated with ratio and proportion.

Curriculum for Excellence (Number 4):

I can use addition, subtraction, multiplication and division when solving problems, making best use of the mental strategies and written skills I have developed.
MNU 103C

Having determined which calculations are needed, I can solve problems involving whole numbers using a range of methods, sharing my approaches and solutions with others.
MNU 203C

Having explored the patterns and relationships in multiplication and division, I can investigate and identify the multiples and factors of numbers.
MTH 207E

I can compare, describe and show number relationships, using appropriate vocabulary and the symbols for equals, not equal to, less than and greater than.
MTH 117R

When a picture or symbol is used to replace a number in a number statement, I can find its value using my knowledge of number facts and explain my thinking to others.
MTH 118R

I can show how quantities that are related can be increased or decreased proportionally and apply this to solve problems in everyday contexts.
MNU 311J

	SHM Resources				Assessment		Other Resources	Date	Comment
Teaching File page	Textbook	Extension Textbook	Pupil Sheet	Home Activity	Check-Up	Topic Assessment			
96–102	22–24		10	5	7				
103–110	25–27		11–12	6	8				
111–114	28–29	E3	13			3a, b			
130–135	32–35	E4–E6	16–17			4a, b			
144–148	36–37			7	9				
149–155	38–43		18–19	8–9	10–11				

Delivering the Curriculum for Excellence

Development planner

Unit	Mathematics 5–14	SHM Topic	Curriculum for Excellence
Number 4 (cont.)		• **Using doubles** – revises doubles of all numbers to 100, multiples of 10 to 1000 and multiples of 100 to 10 000 – introduces a range of strategies for mental multiplication based on doubling and/or halving: – doubling a number ending in 5 when the other number is even and can be halved, for example: – $16 \times 5 \rightarrow 8 \times 10 = 80$ – $25 \times 14 \rightarrow 50 \times 7 = 350$ – halving an even number, for example: – $14 \times 13 \rightarrow 7 \times 13 = 91, 2 \times 91 = 182$ – constructing a 16 times table by doubling the eight times table. • **Written methods of multiplication** – develops an informal written method of multiplication of a 3-digit number by a single digit – uses an expanded vertical recording which leads to the introduction of a standard written method of multiplication of a 3-digit number by a single digit – introduces an informal written method for multiplication of a 2-digit number by a 2-digit number – uses an expanded vertical recording which leads to the introduction of a standard written method of multiplication of a 2-digit number by a 2-digit number – provides opportunities for children to use and apply mental and written methods of multiplication and to use a calculator in a range of problems – uses knowledge and skills in multiplication to solve a variety of extension problems.	I can use addition, subtraction, multiplication and division when solving problems, making best use of the mental strategies and written skills I have developed. **MNU 103C** Having determined which calculations are needed, I can solve problems involving whole numbers using a range of methods, sharing my approaches and solutions with others. **MNU 203C** Having explored the patterns and relationships in multiplication and division, I can investigate and identify the multiples and factors of numbers. **MTH 207E** Having explored more complex number sequences, including well-known named number patterns, I can explain the rule used to generate the sequence, and apply it to extend the pattern. **MTH 221P**
Number 5	**MD/C** • Mentally within the confines of all tables to 10. • In applications in number, measurement and money to £20. **MD/D** • Mentally for whole numbers by single digits • Mentally for 4-digit numbers including decimals by 10 or 100. • Without a calculator for four digits with at most two decimal places by a single digit. • With a calculator for four digits with at most two decimal places by a whole number with two digits • In application in number, measurement and money	**Division** • **Dividing by 2–10, 100, 1000** – revises mental division 2–10 – introduces 'fractional' recording of division (63/7 for 63/7) – extends mental division by 10 and by 100 to – a 4-digit multiple of 100 divided by 10 and by 100 (3600 ÷10; 8200 ÷100) – a 4-digit multiple of 1000 divided by 1000 (5000 ÷1000) – reinforces the link between division and multiplication – includes divisibility tests for the divisions 9, 4, 10, 5, 2 and 100. • **Halving; linking multiplication and division** – consolidates and extends the link between doubling and halving – doubling any number to 100 and halving any even number to 200 – doubling multiple of 10 to 1000 and halving any multiple of 10 to 200, halving any 'even' multiple of 10 to 2000 – doubling a multiple of 100 to 10 000 and halving any 'even' multiple of 100 to 10 000 – includes halving any 3-digit even number – links multiplication and division involving tables facts and larger numbers – uses multiplication to check division. • **Dividing 3-digit numbers, remainders** – introduce division, exact and with remainders, of a 3-digit number by a single digit using: – an informal written method (2-digit quotients) – a standard written method (2- and 3-digit quotients) – introduces rounding to the nearest whole number of a calculator display with several digits after the decimal point – includes dealing with remainders in context. **ASSESSMENT**	I can use addition, subtraction, multiplication and division when solving problems, making best use of the mental strategies and written skills I have developed. **MNU 103C** Having determined which calculations are needed, I can solve problems involving whole numbers using a range of methods, sharing my approaches and solutions with others. **MNU 203C** Having explored the patterns and relationships in multiplication and division, I can investigate and identify the multiples and factors of numbers. **MTH 207E** Having explored more complex number sequences, including well-known named number patterns, I can explain the rule used to generate the sequence, and apply it to extend the pattern. **MTH 221P**
Number 6	**RTN/C** • **Work with** thirds, fifths, eighths, tenths and simple equivalences such as one half = two quarters (practical applications only) **FPR/C** • **Find simple fractions** ($\frac{1}{2}, \frac{1}{5}, \frac{1}{10}$) of quantities involving 1- or 2-digit numbers.	**Fractions** • **Halves, quarters, tenths, thirds and fifths** – revises halves, quarters, tenths, thirds and fifths of a shape – introduces mixed numbers involving these fractions, for example: $2\frac{1}{2}, 3\frac{2}{3}, 4\frac{3}{5}$ – revises finding one half, one quarter and one tenth of a number and introduces finding one third and one fifth of a number.	Having explored fractions by taking part in practical activities, I can show my understanding of: • how a single item can be shared equally • the notation and vocabulary associated with fractions • where simple fractions lie on the number line. **MNU 104H** Through exploring how groups of items can be shared equally, I can find a fraction of an amount by applying my knowledge of division. **MNU 105H** Through taking part in practical activities including use of pictorial representations, I can demonstrate my understanding of simple fractions which are equivalent. **MTH 106H** I have investigated the everyday contexts in which simple fractions, percentages or decimal fractions are used and can carry out the necessary calculations to solve related problems. **MNU 208H** I can show the equivalent forms of simple fractions, decimal fractions and percentages and can choose my preferred form when solving a problem, explaining my choice of method. **MNU 209H** I have investigated how a set of equivalent fractions can be created, understanding the meaning of simplest form, and can apply my knowledge to compare and order the most commonly used fractions. **MTH 210H**

SHM Resources					Assessment		Other Resources	Date	Comment
Teaching File page	Textbook	Extension Textbook	Pupil Sheet	Home Activity	Check-Up	Topic Assessment			
156–160	44		20–21						
161–168	45–47		22–24			5a, b			
176–181	48–51		25–26	10	12				
182–187	52–53				13				
188–191	54–55	E8–E9	29–33		14	6a, b			
222–225	62		38						

Development planner

Unit	Mathematics 5–14	SHM Topic	Curriculum for Excellence	
Number 6 (cont.)		• **Fraction of a shape; equivalence** – revises halves, thirds, quarters, fifths, sixths, eights and tenths of a shape – introduce proper/improper fraction language and the conversion of improper fractions to mixed numbers and vice versa – introduces sevenths, ninths, twelfths and twentieths, including mixed numbers – revises equivalence among fractions and extends this to newly introduced fractions – deals with comparing and ordering fractions with different denominators. **Fraction of a set/quantity** – revises finding a simple fraction of a set ($\frac{1}{2}, \frac{1}{3}, \frac{1}{4}, \frac{1}{5}, \frac{1}{10}$) and a quantity ($\frac{1}{2}, \frac{1}{4}, \frac{1}{10}$) – introduces finding other fractions, numerator = 1, of a set ($\frac{1}{6}, \frac{1}{7}, \frac{1}{8}, \frac{1}{9}$) – includes expressing one quantity as a fraction ($\frac{1}{2}, \frac{1}{4}, \frac{3}{4}$ and tenths) of another, for example: expressing 40p as £$\frac{4}{10}$, 900 m as $\frac{9}{10}$km.	Having explored fractions by taking part in practical activities, I can show my understanding of: • how a single item can be shared equally • the notation and vocabulary associated with fractions • where simple fractions lie on the number line. **MNU 104H** Through exploring how groups of items can be shared equally, I can find a fraction of an amount by applying my knowledge of division. **MNU 105H** Through taking part in practical activities including use of pictorial representations, I can demonstrate my understanding of simple fractions which are equivalent. **MTH 106H** I have investigated the everyday contexts in which simple fractions, percentages or decimal fractions are used and can carry out the necessary calculations to solve related problems. **MNU 208H** I can show the equivalent forms of simple fractions, decimal fractions and percentages and can choose my preferred form when solving a problem, explaining my choice of method. **MNU 209H** I have investigated how a set of equivalent fractions can be created, understanding the meaning of simplest form, and can apply my knowledge to compare and order the most commonly used fractions. **MTH 210H**	
Number 7	**M/C** • **Use coins/notes** of £5 worth or more, including exchange **M/D** • **Use all UK coins/notes** to £20 worth or more, including exchange **RTN/C** • **Work with** decimals to two places when reading/recording money, and using calculator displays.	**Money** • **Amounts using notes and coins** – consolidates counting and laying out money amounts using £10 and £5 notes and all coins – extends conversion from pounds and pence to pence vice versa to 4-digit amounts such as £17.38 ↔ 1738p – introduces rounding to the nearest pound and uses this to estimate the approximate total cost of 2 and 3 items. • **Using mental strategies** – involves finding the total cost of: – 2, 3 or 4 items priced in multiples of 10p, for example, £1.70 + 50p + 30p – 2 or 3 items priced in multiples of 5p, for example, £3.35 + £1.15 – 2 items priced in multiples of 1p, for example, £2.63 + £1.26 – introduces finding change from £20 having spent: – a multiple of 10p, for example £12.40 – a multiple of 5p, for example £7.95 – a multiple of 1p, greater than 50p, for example, £8.76 – provides opportunities for using and applying involving all four operations.	I can use my knowledge of rounding to routinely estimate the answer to a problem, then after calculating, decide if my answer is reasonable, sharing my solution with others. **MNU 201A** I have extended the range of whole numbers I can work with and having explored how decimal fractions are constructed, can explain the link between a digit, its place and its value. **MNU 202B** Having determined which calculations are needed, I can solve problems involving whole numbers using a range of methods, sharing my approaches and solutions with others. **MNU 203C** I have explored the contexts in which problems involving decimal fractions occur and can solve related problems using a variety of methods. **MNU 204C** I can use money to pay for items and can work out how much change I should receive. **MNU 107K** I have investigated how different combinations of coins and notes can be used to pay for goods or be given in change. **MNU 108K** I can manage money, compare costs from different retailers, and determine what I can afford to buy. **MNU 211K**	
Number 8	**AS/C** • Without a calculator, for 4 digits with at most two decimal places • With a calculator, for four digits with at most two decimal places **RTN/D** • **Work with** decimals to two places and equivalences among these in applications in money and measurement	**Decimals** • **Tenths** – introduces decimal notation for tenths in 2- and 3-digit numbers, for example: 0.2, 3.7 and 26.4 – links the decimal notation to the corresponding notation for fractions and mixed numbers, for example: $\frac{2}{10} \rightarrow 0.2$, $3\frac{7}{10} \rightarrow 3.7$ – includes comparing and ordering 2- and 3-digit 1-place decimals and simple sequences – introduces rounding 1-place decimals to the nearest whole number – uses and applies decimal notation in measure and money contexts – introduces the addition and subtraction of 1-place decimals using mental strategies – introduces the standard written method for the addition and subtraction of 1-place decimals and multiplication of 1-place decimals by 2 to 10 as extension activities.	I can use my knowledge of rounding to routinely estimate the answer to a problem, then after calculating, decide if my answer is reasonable, sharing my solution with others. **MNU 201A** I have extended the range of whole numbers I can work with and having explored how decimal fractions are constructed, can explain the link between a digit, its place and its value. **MNU 202B** Having determined which calculations are needed, I can solve problems involving whole numbers using a range of methods, sharing my approaches and solutions with others. **MNU 203C** I have explored the contexts in which problems involving decimal fractions occur and can solve related problems using a variety of methods. **MNU 204C**	
Number 9	**PS/C** • **Work with patterns and relationships** within and among multiplication tables **PS/D** • **Continue and describe more complex relationships**	**Number properties** • **Number properties** – explores number patterns with grids, and investigates further odd and even numbers – introduces using a calculator to generate number sequences – includes multiples of 6, 7, 8, 9 and 11 and common multiples – develops working with factors to include factor pairs and prime numbers – uses factors in multiplication and division – introduces as extension work, square numbers, negative numbers and problem-solving activities.	I can continue and devise more involved repeating patterns or designs, using a variety of media. **MTH 115P** Through exploring number patterns, I can recognise and continue simple number sequences and can explain the rule I've applied. **MTH 116P** Having explored more complex number sequences, including well-known named number patterns, I can explain the rule used to generate the sequence, and apply it to extend the pattern. **MTH 221P** Having explored the patterns and relationships in multiplication and division, I can investigate and identify the multiples and factors of numbers. **MTH 207E** Having determined which calculations are needed, I can solve problems involving whole numbers using a range of methods, sharing my approaches and solutions with others. **MNU 203C** I can show my understanding of how the number line extends to include numbers less than zero and have investigated how these numbers occur and are used. **MNU 206D**	

SHM Resources					Assessment		Other Resources	Date	Comment
Teaching File page	Textbook	Extension Textbook	Pupil Sheet	Home Activity	Check-Up	Topic Assessment			
226–236	63–65		39–42	14	17				
237–238	66	E10		15		8			
202–207	56–57		34–35	11	15				
208–217	58–61		36–37	12	16	7			
248–267	67–73	E11–E13	43–46	16–17	18–19	9a, b			
272–286	74–76	E14–E17	47–52						
Teaching File page	Textbook	Extension Textbook	Pupil Sheet	Home Activity	Check-Up	Topic Assessment			

Delivering the Curriculum for Excellence

Development planner

Unit	Mathematics 5–14	SHM Topic	Curriculum for Excellence
Measure 1	**ME/C** • **Measure in standard units:** – Weight: 1 kg = 1000 g. • **Read scales** on measuring devices to the nearest graduation where the value of an intermediate graduate may be deduced.	**Measure** • **Weight** – consolidates the relationships 1 kg = 1000 g and $\frac{1}{2}$ kg = 500 g and introduces the notation $\frac{1}{4}$ kg = 250 g and $\frac{3}{4}$ kg = 750 g – introduces recording of weights given as kilograms and grams in grams and vice versa: 4 kg 125 g ↔ 4125 g – introduces 200 g, 100 g and 50 g weights and provides estimating and weighing activities using a 2–pan balance – introduces reading scales marked in 10 g, 20 g, 50 g and 100 g divisions – introduces measuring and recording weights to the nearest 100 g – introduces the relationship $\frac{1}{10}$ kg = 100 g – introduces recording weights using decimal notation, for example, 4500 g = 4.5 kg.	I can use my knowledge of the sizes of familiar objects or places to assist me when making an estimate of measure. **MNU 217M** I can use the common units of measure, convert between related units of the metric system and carry out calculations when solving problems. **MNU 218M**
Measure 2	**ME/C** • **Measure in standard units.** • **Estimate length and height** in easily handled standard units: m, $\frac{1}{2}$ m, $\frac{1}{10}$ m, cm. • **Select appropriate measuring devices and units** for length. • **Read scales** on measuring devices to the nearest graduation where the value of an intermediate graduation may be deduced.	**Measure** • **Length** – revises measuring in metres and centimetres – introduces estimating and measuring in lengths to the nearest quarter-metre – introduces recording lengths in decimal form, for example: 1 m 36 cm → 136 cm → 1.36 m – revises measuring and drawing lengths to the nearest half-centimetres – involves choosing appropriate units of length and measuring instruments – introduces the kilometre – includes extension activities which deal with perimeter.	I can use my knowledge of the sizes of familiar objects or places to assist me when making an estimate of measure. **MNU 217M** I can use the common units of measure, convert between related units of the metric system and carry out calculations when solving problems. **MNU 218M** I can explain how different methods can be used to find the perimeter and area of a simple 2D shape or volume of a simple 3D object. **MNU 219M**
Measure 3	**ME/C** • **Measure in standard units:** – volume: litre $\frac{1}{2}$ litre, $\frac{1}{4}$ litre. • **Read scales** on measuring devices to the nearest graduation where the value of an intermediate graduation may be deduced.	**Measure** • **Volume/Capacity** – revises estimating and measuring in litres and half-litres – introduces the quarter-litre and the relationship $\frac{1}{4}$ litre = 250 millilitres, $\frac{3}{4}$ litre = 750 millilitres – uses the notation l and ml when recording volumes in litres and millilitres as millilitres and vice-versa for example: 2 l 400 ml = 2400 ml – revises reading scales marked in 100 ml, 200 ml and 500 ml intervals and extends this to scales marked in 250 ml, 50 ml, 25 ml and 10 ml divisions – integrates different aspects of measure in an extension activity – introduces the cubic centimetre and the notation cm^3 – deals with finding the volume in cubic centimetres of 3D shapes built using cubes.	I can use my knowledge of the sizes of familiar objects or places to assist me when making an estimate of measure. **MNU 217M** I can use the common units of measure, convert between related units of the metric system and carry out calculations when solving problems. **MNU 218M** I can explain how different methods can be used to find the perimeter and area of a simple 2D shape or volume of a simple 3D object. **MNU 219M**
Measure 4	**ME/C** • **Measure in standard units:** – area: shapes composed of rectangles/ squares or irregular shapes using tiles or grids in square centimetres and metres. • **Realise that area can be conserved** when shape changes.	**Measure** • **Area** – revises finding the area of a shape in square centimetres, cm^2 – introduces the formula for finding the area of a rectangle in words – provides practice in finding the approximate areas of irregular shapes – introduces the square metre and the notation m^2.	I can estimate the area of a shape by counting squares or other methods. **MNU 113M** I can explain how different methods can be used to find the perimeter and area of a simple 2D shape or volume of a simple 3D object. **MNU 219M**
Time 1	**T/C** • Use 12–hour times for simple timetables. • Conventions for recording time. • Use calendars.	**Time** • **Reading and writing times** – revises reading the time in 5–minute intervals on analogue and digital displays – revises writing times in 5–minute intervals using 12–hour notation (5 minutes to 8 → 7.55) and am/pm – introduces reading and writing the time in 1-minute intervals on analogue and digital displays using 12–hour notation and am/pm.	I can tell the time using 12 and 24 hour clocks, explain how it impacts on my daily routine and ensure that I am organised and ready for events throughout my day. **MNU 109L**

SHM Resources					Assessment		Other Resources	Date	Comment
Teaching File page	Textbook	Extension Textbook	Pupil Sheet	Home Activity	Check-Up	Topic Assessment			
306–317	83–88		54–57						
290–302	77–82	E18–E19	53						
320–327	89–92		58–59						
330–335	93–94	E20	60–61						
338–341	96		62	18	20				

Delivering the Curriculum for Excellence

Development planner

Unit	Mathematics 5–14	SHM Topic	Curriculum for Excellence	
Time 2	• Conventions for recording time. • Use calendars.	**Time** • **Durations** – revises finding times 5, 10, 15 … 55 minutes before/after given analogue or digital times – revises finding durations in multiples of 5 minutes (to 55 minutes) between given analogue or digital times within the same hour and bridging an hour – introduces finding times 65, 70, 75 … 115 minutes before/after a given analogue or digital time bridging only **one** hour – introduces finding durations in multiples of 5 minutes (greater than 60 minutes) between given analogue or digital times bridging **one** hour – extends finding times before/after given times and durations between analogue or digital times to include 1–minute times within the same hour – includes problems which require the children to use and apply the above – includes estimating and measuring activities involving units of times.	I can tell the time using 12 and 24 hour clocks, explain how it impacts on my daily routine and ensure that I am organised and ready for events throughout my day. **MNU 109L** I have begun to develop a sense of how long tasks take by measuring the time taken to complete a range of activities using a variety of timers. **MNU 111L**	
Shape 1	**RS/C** • **Collect, discuss, make and use 3D shapes.** • Identify 2D shapes within 3D shapes. • Recognise 3D shapes from 2D drawings.	**Shape** • **3D shape** – consolidates recognising, naming, classifying and describing spheres, hemispheres, cubes, cuboids, cones, cylinders, pyramids and prisms – introduces the octahedron – deals with properties associated with these shapes – faces, edges, vertices.	I have explored simple 3D objects and 2D shapes and can identify, name and describe their features using appropriate vocabulary. **MTH 119S** Having explored a range of 3D objects and 2D shapes, I can use mathematical language to describe their properties, and through investigation can discuss where and why particular shapes are used in the environment. **MTH 223S**	
Shape 2	**RS/C** • **Collect, discuss, make and use 2D shapes.**	**Shape** • **2D shape** – revises recognising and describing quadrilaterals and other polygons – develops the language of shape to include parallel, perpendicular and diagonal – introduces work on the properties of equilateral and isosceles triangles.	I have explored simple 3D objects and 2D shapes and can identify, name and describe their features using appropriate vocabulary. **MTH 119S** Having explored a range of 3D objects and 2D shapes, I can use mathematical language to describe their properties, and through investigation can discuss where and why particular shapes are used in the environment. **MTH 223S**	
Shape 3	**S/C** • Find lines of symmetry of shapes drawn on squared grids. • Complete the missing half of a simple symmetrical shape or pattern on a squared grid.	**Shape** • **2D shape** – consolidates work on recognising lines of symmetry in shapes and designs – deals with identifying lines of symmetry in regular and irregular polygons – extends work on completing symmetrical patterns with two lines of symmetry to include more complex patterns – introduces sketching the reflection of a simple shape in a mirror line where not all of the edges of the shape are parallel to or perpendicular to the mirror line – provides practical investigations related to tiling and creating circle patterns.	I have explored symmetry in my own and the wider environment and can create and recognise symmetrical pictures, patterns and shapes. **MTH 123V** I can illustrate the lines of symmetry for a range of 2D shapes and apply my understanding to create and complete symmetrical pictures and patterns. **MTH 231V** I can explore and discuss how and why different shapes fit together and create a tiling pattern with them. **MTH 120S**	
Shape 4	**PM/C** • Describe the main features of a familiar journey or route. • Create paths on squared paper described by instructions such as 'Forward 5, right 90, forward 7, left 90'. **A/C** • Know that a right angle is 90°. • Use 'right, acute, obtuse' to describe angles. • Know that a straight angle is 180°.	**Shape** • **Position, movement and angle** – revises co-ordinates and their notation, for example (2, 3), (5,4) – includes drawing/completing on a co-ordinate grid: – a symmetrical shape – 2D shapes such as triangles, quadrilaterals and other polygons – parallel lines – deals with translating a shape on a squared grid and on a co-ordinate grid. – deals with following and describing pathways on a squared grid using commands such as: **FD 1 BK 3 RT 90 LT 90** – introduces acute and obtuse angles.	I can use my knowledge of the co-ordinate system to plot and describe the location of a point on a grid. **MTH 230U** I can illustrate the lines of symmetry for a range of 2D shapes and apply my understanding to create and complete symmetrical pictures and patterns. **MTH 231V** Through practical activities, which include the use of technology, I have developed my understanding of the link between compass points and angles and can describe, follow and record directions, routes and journeys using appropriate vocabulary. **MTH 228T** I have investigated angles in the environment, and can discuss, describe and classify angles using appropriate mathematical vocabulary. **MTH 226T** I can accurately measure and draw angles using appropriate equipment, applying my skills to problems in context. **MTH 227T**	

SHM Resources					Assessment		Other Resources	Date	Comment
Teaching File page	Textbook	Extension Textbook	Pupil Sheet	Home Activity	Check-Up	Topic Assessment			
342–349	97–101		63–66		21				
370–373	110								
354–361	102–104		67						
362–367	105–109		68–70						
376–395	111–118		71–73						

Delivering the Curriculum for Excellence

Development planner

Unit	Mathematics 5–14	SHM Topic	Curriculum for Excellence	
Information handling 1	**C/D** • **By selecting sources of information** for tasks, including a questionnaire which allows several responses to each question. **O/D** • **By using diagrams or tables.** **D/D** • **By constructing graphs (bar, line, frequency polygon) and pie charts:** – involving simple fractions or decimals – involving continuous data which has been grouped. **I/D** • **From a range of displays and databases by** retrieving information subject to one condition.	**Data handling** • **Interpreting graphs/data** – introduces trend graphs – revises the range and mode of a set of data and introduces the median – introduces the mean of a set of data – introduces compound bar charts and considers how aspects of a set of data may be shown more clearly by a compound bar chart than by a bar chart, and vice versa – introduces, in extension activities, simple pie charts, comparing their effectiveness with that of a compound bar chart and a bar line chart. • **Bar charts with class intervals** – introduces bar charts with class intervals – includes, as an extension activity, survey work with an emphasis on designing questionnaires.	Having discussed the variety of ways and range of media used to present data, I can interpret and draw conclusions from the information displayed, recognising that the presentation may be misleading. **MNU 232W** I have carried out investigations and surveys, devising and using a variety of methods to gather information and have worked with others to collate, organise and communicate the results in an appropriate way. **MNU 233W** I can display data in a clear way using a suitable scale, by choosing appropriately from an extended range of tables, charts, diagrams and graphs, making effective use of technology. **MTH 234X**	
Information handling 2	**C/D** • **By selecting sources of information** for tasks, including a questionnaire which allows several responses to each question. **O/D** • **By using a database or spreadsheet table** with up to three fields defined by pupils. **I/D** • **From a range of displays and databases** by retrieving information subject to one condition.	**Data handling** • **Spreadsheets and databases** – consolidates extracting information presented in tabular form – develops entering data on a simple spreadsheet – deals with interpreting information from a complex database. • **Language of probability** – introduces language associated with the probability of an event occurring including: – likelihood: impossible, unlikely, equally likely, likely and certain – chance: no chance, poor chance, even chance, evens, good chance (and certain) – begins to consider the meaning of fair and unfair.	Having discussed the variety of ways and range of media used to present data, I can interpret and draw conclusions from the information displayed, recognising that the presentation may be misleading. **MNU 232W** I have carried out investigations and surveys, devising and using a variety of methods to gather information and have worked with others to collate, organise and communicate the results in an appropriate way. **MNU 233W** I can display data in a clear way using a suitable scale, by choosing appropriately from an extended range of tables, charts, diagrams and graphs, making effective use of technology. **MTH 234X**	
Number 1	**RTN/D** • **Work with:** – whole numbers up to 100 000 (count, order, read/write) – whole numbers up to a million (read/write only). **RN/D** • **Round any number** to the nearest appropriate whole number, 10 or 100.	**Numbers to millions** • **Number sequences to millions** – consolidates the number sequence to 100 thousands – develops the number sequence to millions (7-/8-digit numbers) – includes finding, in relation to ascending and descending sequences: – the numbers 1, 2, 10, 50, 100, 500, 1000, 10 000, 100 000, 1 0000 000 more/less than a given number – multiples of 10, 100, 1000, 10 000, 100 000, 1 000 000 • **Place value** – introduces place value for numbers up to eight digits – introduces adding and subtracting mentally on a calculator, 10/100/1000/100 000/1 000 000 to and from numbers up to 8 digits – introduces, as an extension activity, adding and subtracting mentally multiples of 1000/10000/100 000/1 000 000 to and from numbers with up to seven digits – deals with: – identifying the larger/smaller number in a pair and largest/smallest number in a set of up to five – ordering up to five non-consecutive numbers – finding the number halfway between a pair of multiples of 1 000 000 or 1 000 000 – reading and writing numbers to millions – revises mental multiplication of a 2-/3-/4-digit number by 10 and a 2-/3-digit number by 100 – introduces mental multiplication of a 2-/3-/4-digit number by 1000 – develops mental division to 10, 100 and 1000 of appropriate powers to 10 to include numbers with up to six digits – introduces work with ancient Egyptian number symbols as an extension activity. • **Estimating and rounding** – revise estimating the position on a number line of – a multiple of 100 on a 0–1000 line – a multiple of 20 on a 0–200 line and extends this to a multiple of 1000 or 2500 on a 0–10 000 line.	I have extended the range of whole numbers I can work with and having explored how decimal fractions are constructed, can explain the link between a digit, its place and its value. **MNU 202B** I can use my knowledge of rounding to routinely estimate the answer to a problem, then after calculating, decide if my answer is reasonable, sharing my solution with others. **MNU 201A** Having determined which calculations are needed, I can solve problems involving whole numbers using a range of methods, sharing my approaches and solutions with others. **MNU 203C** I have discussed the important part that numbers play in the world and explored a variety of systems that have been used by civilisations throughout history to record numbers. **MTH 114N**	

SHM Resources					Assessment		Other Resources	Date	Comment
Teaching File page	Textbook	Extension Textbook	Pupil Sheet	Home Activity	Check-Up	Topic Assessment			
364–371	113–117	E20–E21							
372–375	118	E22	56–57						
378–382	119–122		58						
383–386	123					15			
42–46	1		1–3	1	1				
47–59	2–6	E1–E2	4–8	2–4	2				
60–66	7–8		9		3				

Delivering the Curriculum for Excellence

Development planner

Unit	Mathematics 5–14	SHM Topic		
Number 1 (cont.)		– includes problem solving activities which involve estimating quantities – introduces rounding to a 5-/6-digit number to the nearest 1000/100 and a 7-/8-digit number to the nearest million.		
Number 2	**AS/D** • **Add**: – mentally for 2-digit whole numbers, beyond in some cases, involving multiples of 10 or 100 – without a calculator, for four digits with at most two decimal places – with a calculator, for four digits with at most two decimal places.	**Addition** • **Mental addition involving 2-/3-digit numbers** – consolidates and develops strategies for mental addition of several 2-digit numbers – introduces finding an approximate total based on rounding 3-digit numbers to the nearest hundred/nearest 10 – consolidates mental addition of 3-digit numbers, bridging a multiple of 10, for example: 428 + 267, 149 + □ = 682 – introduces mental addition of 3-digit numbers – bridging a multiple of 100, for example: 493 + 272, 193 + □ = 753 – bridging 1000, for example: 635 + 853, 650 + □ = 1170 – introduces the use of a doubling strategy for mental addition of two numbers close to and on either side of the same multiple of 100, for example: 421 + 387. • **Addition involving numbers with up to four digits** – introduces mental addition of 4-digit multiples of 100, for example: 6100 + 2300, 4400 + 3800, 6500 + 7600 – introduces mental addition of 4-digit numbers with no bridging (2634 + 2352) then bridging a multiple of 10 only (4564 + 2107) – further develops the use of a standard written method of addition of: – two numbers with four digits – several numbers with different numbers of digits to include totals greater than 10 000.	I can use my knowledge of rounding to routinely estimate the answer to a problem, then after calculating, decide if my answer is reasonable, sharing my solution with others. **MNU 201A** I have extended the range of whole numbers I can work with and having explored how decimal fractions are constructed, can explain the link between a digit, its place and its value. **MNU 202B** Having determined which calculations are needed, I can solve problems involving whole numbers using a range of methods, sharing my approaches and solutions with others. **MNU 203C** I can compare, describe and show number relationships, using appropriate vocabulary and the symbols for equals, not equal to, less than and greater than. **MTH 117R** When a picture or symbol is used to replace a number in a number statement, I can find its value using my knowledge of number facts and explain my thinking to others. **MTH 118R**	
Number 3	**AS/D** • **Subtract**: – mentally for 2-digit whole numbers, beyond in some cases, involving multiples of 10 or 100 – without a calculator, for four digits with at most two decimal places – with a calculator, for four digits with at most two decimal places.	**Subtraction** • **Mental subtraction involving 3-digit numbers** – revises mental subtraction of a 2-digit number from a 3-digit number – revises mental subtraction involving three-digit multiples of 10 (550 – 260) and introduces mental subtraction, bridging a multiple of 100: – of a 3-digit multiple of 10 from any 3-digit number (768 – 380) – of any 3-digit number from a 3-digit multiple of 10 (650 – 464) – introduces mental subtraction of 3-digit numbers, bridging a multiple of 10 (862 – 349). • **Subtraction involving numbers with four or more digits** – revises mental subtraction from a 4-digit multiple of 100 and then from any 4-digit number of a 3-digit multiple of 100 (4300 – 600 → 8197 – 500) – introduces mental subtraction from a 4-digit multiple of 100 of: – another 4-digit multiple of 100, bridging a multiple of 100 (7300 – 6800) – a 4-digit multiple of 50, not bridging a multiple of 1000 (8400 – 5150) – introduces mental subtraction from a multiple of 1000 of any 3-/4- digit number (2000 – 754, 8000 – 2785) – introduces mental subtraction from a 4-digit number of a 3-/4-digit number: – with no bridging (3956 – 703, 7568 – 1352) – bridging a multiple of 10 only (2894 – 745, 6722 – 2607) – consolidates a standard written method of subtraction and includes subtractions involving: – 4-digit numbers (9672 – 4883) – different numbers of digits (3215 – 76) – provides opportunities to use and apply skills in mental subtraction and in using a calculator – provides an extension activity dealing with using a calculator for **addition and subtraction** of numbers with more than four digits.	I can use my knowledge of rounding to routinely estimate the answer to a problem, then after calculating, decide if my answer is reasonable, sharing my solution with others. **MNU 201A** I have extended the range of whole numbers I can work with and having explored how decimal fractions are constructed, can explain the link between a digit, its place and its value. **MNU 202B** Having determined which calculations are needed, I can solve problems involving whole numbers using a range of methods, sharing my approaches and solutions with others. **MNU 203C**	
Number 4	**MD/D** • **Multiply**: – mentally for whole numbers by single digits – mentally for 4-digit numbers including decimals by 10 or 100 – without a calculator for four digits with at most two decimal places by a single digit – with a calculator for four digits with at most two decimal places by a whole number with 2 digits in applications in number, measurement and money.	**Multiplication** • **Mental multiplication** – revises multiplication facts for 2 to 10 and finding mentally products involving multiples of 10/100/1000, for example: 20×9, 400×8, 6×3000, 40×700 – revises multiplying a 2-digit number by a single digit with bridging, for example: 6×37 → $6 \times 30 = 180$, $6 \times 7 = 42$ → $180 + 42 = 222$ – revises a range of mental multiplication strategies based on the distributive property and/or doubling/halving: – using doubling to build up a 'multiplication table' for a 2-digit number.	I can use my knowledge of rounding to routinely estimate the answer to a problem, then after calculating, decide if my answer is reasonable, sharing my solution with others. **MNU 201A** I have extended the range of whole numbers I can work with and having explored how decimal fractions are constructed, can explain the link between a digit, its place and its value. **MNU 202B** Having determined which calculations are needed, I can solve problems involving whole numbers using a range of methods, sharing my approaches and solutions with others. **MNU 203C** Having explored the patterns and relationships in multiplication and division, I can investigate and identify the multiples and factors of numbers. **MTH 207E**	

SHM Resources					Assessment		Other Resources	Date	Comment
Teaching File page	Textbook	Extension Textbook	Pupil Sheet	Home Activity	Check-Up	Topic Assessment			
						57			
70–77	9–13		10	5	4				
78–83	14–18			6	5	1a, b			
90–95	19–21			7	6				
96–103	22–24	3	11–12	8	7	2a, b			
112–122	25–29		13–15	9–10	8–9				

Delivering the Curriculum for Excellence

Development planner

Unit	Mathematics 5–14	SHM Topic	Curriculum for Excellence
Number 4 (cont.)		– doubling a number ending in 5 and halving the other number, for example: – $15 \times 16 \to 30 \times 80 = 240$ – halving an even teens number and doubling another number, for example: – $14 \times 23 \to 7 \times 46 = 322$ – introduces mental multiplication strategies based on multiplying by 100, for example: – multiplying by 50 by multiplying by 100 then halving: $50 \times 23 \to 100 \times 23\ (2300) \to$ half of $2300 = 1150$ – multiplying by 25 by multiplying by 100, halving and then halving again: $25 \times 32 \to 100 \times 32\ (3200) \to$ half of $3200 \to$ half of $1600 = 800$ – introduces using factors as strategy for multiplying a pair of 2-digit numbers – uses and applies knowledge of multiplication facts and strategies to solve number problems. • **Written methods, calculator** – develops an informal written method for multiplication of a 1-digit number by a single digit – uses an expanded vertical recording which leads to the further development of a standard written method of multiplication of a 4-digit number by a single digit – consolidates informal and written methods for multiplication of a 2-digit number by a 2-digit number – provides opportunities for using and applying knowledge and skills in multiplication – includes multiplication of 3-/4-digit numbers by a 2-digit number using a calculator.	I can use my knowledge of rounding to routinely estimate the answer to a problem, then after calculating, decide if my answer is reasonable, sharing my solution with others. **MNU 201A** I have extended the range of whole numbers I can work with and having explored how decimal fractions are constructed, can explain the link between a digit, its place and its value. **MNU 202B** Having determined which calculations are needed, I can solve problems involving whole numbers using a range of methods, sharing my approaches and solutions with others. **MNU 203C** Having explored the patterns and relationships in multiplication and division, I can investigate and identify the multiples and factors of numbers. **MTH 207E**
Number 5	**MD/D** • **Divide:** – mentally for whole numbers by single digits. – mentally for four-digit numbers including decimals by 10 or 100. – without a calculator for digits with at most two decimal places by a single digit. – with a calculator for four digits with at most two decimal places by a whole number with two digits. – in applications in number, measure and money.	**Division** • **Mental division** – revises mental division based on tables facts, including remainders – consolidates and develops work on finding mentally half of – any 3-digit even number – 4-digit multiples of 10 to 2000 (examples with an even tens digit only i.e. finding half of 1940 but not 1730) – 4-digit multiples of 100 (examples of an even hundred digit only i.e. finding half of 5800 but not 7500) – introduces mental strategies for division by a single digit of – certain 3-digit multiples of 10 ($320 \div 8 \to 32 \div 8 = 4$, $320 \div 8 = \mathbf{40}$) – numbers just beyond the extent of the tables ($52 \div 4 \to 52 = 40 + 12$, $40 \div 4 = 10$, $12 \div 4 = 3$, so $52 \div 4 = \mathbf{13}$ $(10 + 3)$) – certain 3-digit numbers, based on tables facts ($763 \div 7 \to 700 \div 7 = 100$, $63 \div 7 = 9$, $763 \div 7 = \mathbf{109}$) – includes, in extension activity, division word problems involving one million. • **Written division, calculator** – revises and develops division, exact and with remainders, of a 3-digit number by a single digit using: – a short standard written method based on place value sharing – an alternative written method involving repeated subtraction of multiples of the divisor (2- and 3-digit quotients) – introduces (using the **school's choice** of written method) division, exact and with remainders, of a 4-digit number by a single digit (3- and 4-digit quotients).	I can use my knowledge of rounding to routinely estimate the answer to a problem, then after calculating, decide if my answer is reasonable, sharing my solution with others. **MNU 201A** I have extended the range of whole numbers I can work with and having explored how decimal fractions are constructed, can explain the link between a digit, its place and its value. **MNU 202B** Having determined which calculations are needed, I can solve problems involving whole numbers using a range of methods, sharing my approaches and solutions with others. **MNU 203C** Having explored the patterns and relationships in multiplication and division, I can investigate and identify the multiples and factors of numbers. **MTH 207E**
Number 6	**PS/D** • **Continue and describe more complex sequences.** **FE/D** • **Recognise and explain simple relationships:** – between two sets of numbers or objects. **MD/D** • **Multiply and divide:** – mentally for whole numbers by single digits.	**Number properties** • **Number sequences and patterns** – consolidates continuing number sequence and using 'rules' to describe or generate number sequences, including sequences which: – increase or decrease in single-digit steps and in steps of 11/15/19/21/25 – involve doubling/halving – consolidates language (square, squared) and notation ($25 = 5 \times 5 = 5^2$) associated with square numbers and introduces, in a follow-up activity, the idea of a square root – explores types of numbers in investigations which lead to generalisations about products of odd/even numbers, for example, 'the product of two odd/even numbers is an odd/even number' – investigates, in extension activities: – ordering and addition and subtraction of negative numbers, in the context of temperature – identifying and continuing number patterns.	I can continue and devise more involved repeating patterns or designs, using a variety of media. **MTH 115P** Through exploring number patterns, I can recognise and continue simple number sequences and can explain the rule I've applied. **MTH 116P** Having explored more complex number sequences, including well-known named number patterns, I can explain the rule used to generate the sequence, and apply it to extend the pattern. **MTH 221P** I can compare, describe and show number relationships, using appropriate vocabulary and the symbols for equals, not equal to, less than and greater than. **MTH 117R** When a picture or symbol is used to replace a number in a number statement, I can find its value using my knowledge of number facts and explain my thinking to others. **MTH 118R** I can apply my knowledge of number facts to solve problems where an unknown value is represented by a symbol or letter. **MTH 222R** I can show my understanding of how the number line extends to include numbers less than zero and have investigated how these numbers occur and are used. **MNU 206D** Having explored the patterns and relationships in multiplication and division, I can investigate and identify the multiples and factors of numbers. **MTH 207E**

SHM Resources					Assessment		Other Resources	Date	Comment
Teaching File page	Textbook	Extension Textbook	Pupil Sheet	Home Activity	Check-Up	Topic Assessment			
123–128	30–32		16–17			3a, b			
134–140	33–34		18–19	12–13	10				
141–145	35–36		20–22		11	4a, b			
152–163	37–39	4–6	23–24	14					

Development planner

Unit	Mathematics 5–14	SHM Topic	Curriculum for Excellence	
Number 6 (cont.)		• **Divisibility, multiples and factors, word formulae** – revises methods (which involve consideration of the last digit/s or the digit sum) of 'testing' numbers, without dividing, for exact divisibility by 2, 3, 4, 5, 9, 10 and 100 – introduces methods (which involve a 'two-step' process) of testing numbers without dividing, for divisibility by 4, 6 and 8, for example: 168 is exactly divisible by 8 because half of it, 84 (168 ÷ 2), is exactly divisible by 4 – consolidates multiples and common multiples and introduces the idea of smallest/lowest common multiple for a pair of numbers, for example: 3 and 4, 10 and 15, 4 and 16 – consolidates factors, including finding all the factor pairs of a number and listing all its factors, and by investigating sets of numbers leads to the discoveries that: – a square number has an odd number of factors – a prime number has only two factors, itself and 1 – introduces finding the 'rule' or word formula to describe a relationship between two sets of numbers – uses and applies knowledge of number properties to solve a range of problems and puzzles.	I can use my knowledge of rounding to routinely estimate the answer to a problem, then after calculating, decide if my answer is reasonable, sharing my solution with others. **MNU 201A** I have extended the range of whole numbers I can work with and having explored how decimal fractions are constructed, can explain the link between a digit, its place and its value. **MNU 202B** Having determined which calculations are needed, I can solve problems involving whole numbers using a range of methods, sharing my approaches and solutions with others. **MNU 203C** Having explored the patterns and relationships in multiplication and division, I can investigate and identify the multiples and factors of numbers. **MTH 207E** Having explored more complex number sequences, including well-known named number patterns, I can explain the rule used to generate the sequence, and apply it to extend the pattern. **MTH 221P**	
Number 7	**RTN/D** • **Work with:** • Fractions (all previous plus twentieths, fiftieths, hundredths) and equivalences among these and decimals (in applications). **FPR/D** • **Work with fractions and percentages** • Find simple factions ($\frac{1}{2}$, $\frac{3}{4}$, $\frac{3}{5}$, $\frac{60}{100}$) of quantities involving at most four digits (easy examples only).	**Fractions** • **Equivalence** – consolidates proper/improper fraction language and the conversion of mixed numbers to improper fractions and vice versa – revises forming equivalent fractions by multiplying numerator and denominator and extends this to any fraction – introduces forming equivalent fractions by dividing numerator and denominator and associated language (simplifying, simplest form) – deals with comparing and ordering fractions with different denominators – includes, as an extension activity, informal addition and subtraction of fractions and mixed numbers involving quarters, halves and three quarters. • **Fraction of a set/quantity; hundredths** – consolidates finding a fraction, numerator = 1, of a set – introduces fractional notation, including mixed numbers, for hundredths – introduces finding 'any' fraction, numerator ≥ 1, of a set/quantity, for example: – $\frac{2}{7}$ of 21, $\frac{7}{100}$ of 500, $\frac{4}{5}$ of 45 m, $\frac{3}{8}$ of 2 kg, including finding several hundredths of £1, 1 m, 1 km, 1 kg and 1 ℓ, for example: $\frac{9}{100}$ of 1 km, $\frac{6}{100}$ ℓ – develops expressing one quantity as a fraction of another to include: – for tenths, quantities greater than £1 and 1 m, for example: expressing £1.30 as £1 $\frac{3}{10}$, 180 cm as 1 $\frac{8}{10}$ m – pence, cm, m, g and ml as hundredths of £1, 1 m, 1 km, 1 kg and 1 ℓ respectively, for example: expressing 20 m as $\frac{2}{100}$ km, 70 ml as $\frac{7}{100}$ ℓ.	Having explored fractions by taking part in practical activities, I can show my understanding of: • how a single item can be shared equally • the notation and vocabulary associated with fractions • where simple fractions lie on the number line. **MNU 104H** Through exploring how groups of items can be shared equally, I can find a fraction of an amount by applying my knowledge of division. **MNU 105H** Through taking part in practical activities including use of pictorial representations, I can demonstrate my understanding of simple fractions which are equivalent. **MTH 106H** I have investigated the everyday contexts in which simple fractions, percentages or decimal fractions are used and can carry out the necessary calculations to solve related problems. **MNU 208H** I can show the equivalent forms of simple fractions, decimal fractions and percentages and can choose my preferred form when solving a problem, explaining my choice of method. **MNU 209H** I have investigated how a set of equivalent fractions can be created, understanding the meaning of simplest form, and can apply my knowledge to compare and order the most commonly used fractions. **MTH 210H**	
Number 8	**RTN/D** • **Work with** fractions (all previous plus twentieths, fiftieths, hundredths) and equivalences among these and decimals (in applications). **AS/D** • Without a calculator, for four digits with at most two decimal places. • With a calculator, for four digits with at most two decimal places.	**Decimals** • **Tenths** – revises decimal notation for tenths in 2– and 3-digit numbers, for example: 2.6, 27.9 – links the decimal notation to the corresponding notation for fractions and mixed numbers, for example: $\frac{3}{10}$ → 0.3, 2 $\frac{7}{10}$ → 2.7 – includes work on place value, sequences, comparing and ordering 2–and 3-digit decimals – deals with mental methods for adding and subtracting 2-digit decimals, for example: 1.7 + 5.6, 7.2 – 3.9 – uses standard written methods for adding and subtracting 3-digit decimals, for example: 57.5 + 38.4, 93.9 – 77.2 – deals with mental multiplication of a 2-digit number with one decimal place by a single digit – introduces multiplication of a 3-digit number with one decimal place by a single digit, using informal and standard written methods – introduces mental division of a 1- or 2-digit number by 10 – introduces mental division of a 2-digit number with one decimal place by a single digit – introduces division of a 3-digit number with one decimal place by a single digit using a standard written method.	I can use my knowledge of rounding to routinely estimate the answer to a problem, then after calculating, decide if my answer is reasonable, sharing my solution with others. **MNU 201A** I have extended the range of whole numbers I can work with and having explored how decimal fractions are constructed, can explain the link between a digit, its place and its value. **MNU 202B** Having determined which calculations are needed, I can solve problems involving whole numbers using a range of methods, sharing my approaches and solutions with others. **MNU 203C** I have explored the contexts in which problems involving decimal fractions occur and can solve related problems using a variety of methods. **MNU 204C** Having explored the need for rules for the order of operations in number calculations, I can apply them correctly when solving simple problems. **MTH 205C**	

Teaching File page	Textbook	Extension Textbook	Pupil Sheet	Home Activity	Check-Up	Topic Assessment	Other Resources	Date	Comment
	SHM Resources				Assessment				
164–175	40–47		25			5a, b			
182–190	48–50	10	26–28	15	12				
191–198	51–53		29	16	13	6a, b			
208–220	54–60		30–35	17–18	14–16				

Delivering the Curriculum for Excellence

Development planner

Unit	Mathematics 5–14	SHM Topic	Curriculum for Excellence
Number 9	**RTN/D** • **Work with** decimals to two place and equivalences among these in applications in money and measurement **AS/D** • Without a calculator, for four digits with at most two decimal places • With a calculator, for four digits with at most two decimal places.	**Decimals** • **Hundredths** – introduces decimal notation for hundredths in numbers with up to four digits – includes linking decimal notation to the corresponding notation for fractions and mixed numbers, for example: $0.39 \rightarrow \frac{39}{100}$, $46.2 \rightarrow 46\frac{2}{10} \rightarrow 46\frac{1}{5}$ or $46\frac{20}{100}$ – deals with place value for 1- and 2-place decimals with up to four digits and includes recognising: – the larger/smaller decimal in a pair of numbers – the largest/smallest decimal in a set of three numbers – the 2-place decimal before/after/between – includes ordering sets of up to six 1-/2-place decimals with up to three digits, as sample sequences – introduces addition and subtraction of 2-place decimals using mental strategies for 3-digit numbers and standard written methods (for 3-/4-digit numbers) – uses and applies decimal notation in the contexts of length, for example 3.67 m and money, for example: £12.02 – introduces mental multiplication of a 2-/3-digit number with one/two decimal places by 10, 100 or a multiple of 10 – includes using and applying calculator skills to solve problems involving decimals – introduces, in extension activities, multiplication and division of a 3-digit number with one/two decimal places using standard written methods.	I can use my knowledge of rounding to routinely estimate the answer to a problem, then after calculating, decide if my answer is reasonable, sharing my solution with others. **MNU 201A** I have extended the range of whole numbers I can work with and having explored how decimal fractions are constructed, can explain the link between a digit, its place and its value. **MNU 202B** Having determined which calculations are needed, I can solve problems involving whole numbers using a range of methods, sharing my approaches and solutions with others. **MNU 203C** I have explored the contexts in which problems involving decimal fractions occur and can solve related problems using a variety of methods. **MNU 204C** Having explored the need for rules for the order of operations in number calculations, I can apply them correctly when solving simple problems. **MTH 205C**
Number 10	**RTN/D** • **Work with**: percentages, decimals to two places and equivalences among these in applications in money and measurement **FPR/D** • **Work with fractions and percentages:** – find simple fractions ($\frac{1}{7}$, $\frac{3}{4}$, $\frac{3}{5}$, $\frac{60}{100}$) of quantities involving at most four digits (easy examples only).	**Percentages** • **Percentages** – introduces the concept of percentages – introduces the links between fractions, decimals, and percentages, for example: $\frac{20}{100} = 0.20 = 20\%$ – deals with finding the percentage of a shape which is shaded – includes finding a percentage of a set/quantity.	I can use my knowledge of rounding to routinely estimate the answer to a problem, then after calculating, decide if my answer is reasonable, sharing my solution with others. **MNU 201A** I have extended the range of whole numbers I can work with and having explored how decimal fractions are constructed, can explain the link between a digit, its place and its value. **MNU 202B** Having determined which calculations are needed, I can solve problems involving whole numbers using a range of methods, sharing my approaches and solutions with others. **MNU 203C** I have explored the contexts in which problems involving decimal fractions occur and can solve related problems using a variety of methods. **MNU 204C** I have investigated the everyday contexts in which simple fractions, percentages or decimal fractions are used and can carry out the necessary calculations to solve related problems. **MNU 208H** I can show the equivalent forms of simple fractions, decimal fractions and percentages and can choose my preferred form when solving a problem, explaining my choice of method. **MNU 209H** I have investigated how a set of equivalent fractions can be created, understanding the meaning of simplest form, and can apply my knowledge to compare and order the most commonly used fractions. **MTH 210H**
Measure 1	**T/D** • **Work with time:** – use 24–hour times and equate with 12-hour times.	**Time** • **Reading and writing times** – revises reading the time in 1-minute intervals on analogue and digital displays – revises writing times in 1-minute intervals using 12-hour notation and am/pm – introduces reading and writing the time in 1 minute intervals using 24-hour notation.	I can tell the time using 12 and 24 hour clocks, explain how it impacts on my daily routine and ensure that I am organised and ready for events throughout my day. **MNU 109L**
Measure 2	**T/D** • **Work with time:** – use 24–hour times and equate with 12-hour times – calculate duration in hours/minutes, mentally if possible – time activities in seconds with a stopwatch – calculate speeds.	**Time** • **Duration, seconds** – introduces finding times in multiples of 5 minutes before/after times displayed in 24-hour notation: – bridging 1 hour, for example: 25 min after 08:55 – bridging more than an hour, for example 1 h and 50 min after 22:30 – introduces finding durations in multiples of 5 minutes between digital times given in 24-hour notation, bridging 1 hour/more than 1 hour – includes problems which require the children to use and apply the above – deals with estimating and measuring activities involving units of time – deals with seconds and using and reading stop clocks in practical estimating and measuring activities – introduces *rate* as an extension.	I can tell the time using 12 and 24 hour clocks, explain how it impacts on my daily routine and ensure that I am organised and ready for events throughout my day. **MNU 109L** I have begun to develop a sense of how long tasks take by measuring the time taken to complete a range of activities using a variety of timers. **MNU 111L** I can use and interpret electronic and paper-based timetables and schedules to plan events and activities, and make time calculations as part of my planning **MNU 214L** I can carry out practical tasks and investigations involving timed events and can explain which unit of time would be most appropriate to use. **MNU 215L**
Measure 3	**ME/D** • **Measure in standard units:** – length, small lengths in millimetres; large lengths like buildings in metres. • **Recognise** when kilometres are appropriate.	**Measure** • **Length** – consolidates estimating and measuring in metres and centimetres – includes activities involving: – the millimetre and the relationship 10 mm = 1 cm – the kilometre and the relationship with metres, for example: 1 km = 1000 m, $\frac{1}{10}$ km = 100 m and $\frac{3}{4}$ km = 750 m – involves using practical skills to choose appropriate units and measuring devices and applying knowledge to solve problems – introduces finding the perimeters of shapes by adding the lengths and their sides and introduces word formulae to find the perimeter of a rectangle and a regular polygon.	I can use the common units of measure, convert between related units of the metric system and carry out calculations when solving problems. **MNU 218M** I can explain how different methods can be used to find the perimeter and area of a simple 2D shape or volume of a simple 3D object. **MNU 219M**

SHM Resources					Assessment		Other Resources	Date	Comment
Teaching File page	Textbook	Extension Textbook	Pupil Sheet	Home Activity	Check-Up	Topic Assessment			
221–232	61–68	8–9	36–37	19–21		7a, b			
240–249	69–72	11	38–39	22		8a, b			
256–260	73–74		41	23					
261–266	75–78	12–13	42–43			9			
272–280	79–84		44			10			

Delivering the Curriculum for Excellence

Development planner

Unit	Mathematics 5–14	SHM Topic	Curriculum for Excellence	
Measure 4	**ME/D** • **Measure in standard units:** – weight: extended range of articles, for example, own weight. • **Estimate** small weights in easily handled units. • **Select appropriate measuring devices and units** for weight.	**Measure** • **Weight** – revises reading scales marked in 10 g, 20 g, 50 g, and 100 g divisions – introduces reading and recording weights to the nearest mark on scales showing tenths of a kilogram in decimal form (0.1 kg) and in multiples of 100 g – revises recording in grams weights given in kilograms and grams and vice versa, for example: 4 kg 125 g → 4125 g, 3697 g → 3 kg 697 g – uses and applies knowledge of weight in practical contexts.	I can use the common units of measure, convert between related units of the metric system and carry out calculations when solving problems. **MNU 218M**	
Measure 5	**ME/C** • **Measure in standard units:** – volume, accuracy extended to small containers in millilitres; 1 ℓ = 1000 ml – temperature. • **Estimate** small volumes in easily handled units. • **Be aware of common imperial units** in appropriate practical applications.	**Measure** • **Volume/Capacity** – revises the relationship 1 ℓ = 1000 ml, $\frac{1}{2}$ ℓ = 500 ml and $\frac{3}{4}$ ℓ = 750 ml – introduces the tenth of a litre and the relationship $\frac{1}{10}$ ℓ = 100 ml, $\frac{2}{10}$ ℓ = 200 ml – revises reading scales marked in 100 ml, 50 ml, 20 ml and 10 ml divisions – provides practical estimating and measuring activities and includes solving problems in non-practical situations – introduces, in a problem solving activity, the relationship 1 cm³ has the same volume as 1 ml. • **Mixed measure** – deals with commonly used imperial units – the mile, pint and gallon – and their approximate metric equivalents – includes problems involving weight, volume/capacity, temperature and time.	I can use my knowledge of the sizes of familiar objects or places to assist me when making an estimate of measure. **MNU 217M** I can use the common units of measure, convert between related units of the metric system and carry out calculations when solving problems. **MNU 218M** I can explain how different methods can be used to find the perimeter and area of a simple 2D shape or volume of a simple 3D object. **MNU 219M**	
Measure 6	**ME/C** • **Measure in standard units:** – area: right-angled triangles on cm squared grids. • **Estimate** small areas in easily handled standard units.	**Measure** • **Area** – revises using the formula in words for finding the area of a rectangle and introduces expressing this in letters as A=l×b – applies the formula, A=l×b, to rectangles with side lengths in half centimetres – introduces finding the approximate area of squares and rectangles to the nearest cm² – provides practice in finding the approximate area of irregular shapes – develops finding the area, in cm² and m², of a simple composite shape sub-divided into rectangles – introduces finding the area of a right-angled triangle drawn on a squared grid – consolidates a counting method for finding the area of an irregular shape drawn on a dotty grid in square units and half-square units – includes problems which require the children to use and apply the above – develops, in an extension activity, methods for **calculating** the area of a right-angled triangle.	I can estimate the area of a shape by counting squares or other methods. **MNU 113M** I can use my knowledge of the sizes of familiar objects or places to assist me when making an estimate of measure. **MNU 217M** I can use the common units of measure, convert between related units of the metric system and carry out calculations when solving problems. **MNU 218M** I can explain how different methods can be used to find the perimeter and area of a simple 2D shape or volume of a simple 3D object. **MNU 219M**	
Shape 1	**S/D** • **Work with symmetry:** – identify and draw lines of symmetry, generally up to 4 – create symmetrical shapes.	**2D shape** • **Line symmetry** – revises reflecting simple shapes in mirror lines in a variety of orientations – develops sketching the reflection of a shape in one line of symmetry where at least two sides of the shape are not parallel to or perpendicular to the line of symmetry – introduces reflecting straight and then curved lines to create patterns with both horizontal and vertical lines of symmetry.	I have explored symmetry in my own and the wider environment and can create and recognise symmetrical pictures, patterns and shapes. **MTH 123V** I can illustrate the lines of symmetry for a range of 2D shapes and apply my understanding to create and complete symmetrical pictures and patterns. **MTH 231V**	
Shape 2	**RS/D** • **Collect, discuss, make and use 2D shapes:** – discuss 2D shapes referring to sides, diagonals, angles – recognise pentagon, hexagon – identify and name equilateral and isosceles triangles – extend shape vocabulary to radius, diameter, circumference – create or copy a tiling using a shape template – use the rigidity property of triangle in model-making.	**2D shape** • **Properties, puzzles and patterns** – consolidates naming and describing triangles, quadrilaterals and other polygons using side and angle properties – revises equilateral and isosceles triangles and introduces scalene and right-angled triangles – introduces the parallelogram, rhombus, trapezium and kite, extending the language of shape to include the terms adjacent sides, opposite sides/angles – introduces further work on tiling patterns drawn on squared and isometric dotty grids – introduces using compasses to draw a variety of circle designs – includes, in extension activities, investigations involving: – the rigidity property of the triangle – the 7–piece Tangram – drawing pursuit curve patterns in a range of triangles, quadrilaterals and regular polygons – exploring number patterns arising when intersecting straight lines are drawn between points.	Having explored a range of 3D objects and 2D shapes, I can use mathematical language to describe their properties, and through investigation can discuss where and why particular shapes are used in the environment. **MTH 223S** I can explore and discuss how and why different shapes fit together and create a tiling pattern with them. **MTH 120S** I can draw 2D shapes and make representations of 3D objects using an appropriate range of methods and efficient use of resources. **MTH 225S**	

SHM Resources					Assessment		Other Resources	Date	Comment
Teaching File page	Textbook	Extension Textbook	Pupil Sheet	Home Activity	Check-Up	Topic Assessment			
284–288	85–86		45–46			11			
292–296	87–90		47–48						
297–299	91–92								
302–312	93–97	E19	49–50						
320–326	98		51–52						
327–334	99–104	E14–E18	53–54			13a, b			

Delivering the Curriculum for Excellence

Development planner

Unit	Mathematics 5–14	SHM Topic	Curriculum for Excellence	
Shape 3	**RS/D** • **Collect, discuss, make and use 3D shapes:** – discuss 3D shapes referring to faces, edges, vertices – make 3D models, solid or skeletal, including using nets: cube and cuboid only.	**3D shape** • **3D shape** – explores possible nets of a cube – includes identifying shapes from their nets – introduces finding the total surface area of various 3D shapes – consolidates and develops work on interpreting 2D representation of 3D shapes composed of linking cubes.	Through practical activities, I can show my understanding of the relationship between 3D objects and their nets. **MTH 224S** Having explored a range of 3D objects and 2D shapes, I can use mathematical language to describe their properties, and through investigation can discuss where and why particular shapes are used in the environment. **MTH 223S** I can explain how different methods can be used to find the perimeter and area of a simple 2D shape or volume of a simple 3D object. **MNU 219M** I can draw 2D shapes and make representations of 3D objects using an appropriate range of methods and efficient use of resources. **MTH 225S**	
Shape 4	**PM/D** • **Discuss position and movement:** – give directions for a route or journey – use an 8-point compass rose – use a co-ordinate system to locate a point on a grid – create patterns by rotating a shape. **A/D** • **Angles:** – draw, copy and measure angles accurately within 5 degrees – use standard notation, 060°, 150°, 300°, to express bearings.	**Shape** • **Position, movement and angle** – uses the 8-point compass and a 12–point dial respectively to consolidate in multiples of 45° and 30° – introduces bearings and the associated 3-figure notation, for example 090° – introduces drawing, on a square grid, shapes which have been rotated through 90°, 180° and 270° and includes locating the positions of the vertices of shapes on a co-ordinate grid, before and after rotation – introduces estimating and measuring angles to the nearest 5°.	I have investigated angles in the environment, and can discuss, describe and classify angles using appropriate mathematical vocabulary. **MTH 226T** I can accurately measure and draw angles using appropriate equipment, applying my skills to problems in context. **MTH 227T** Through practical activities, which include the use of technology, I have developed my understanding of the link between compass points and angles and can describe, follow and record directions, routes and journeys using appropriate vocabulary. **MTH 228T** Having investigated where, why and how scale is used and expressed, I can apply my understanding to interpret simple models, maps and plans. **MTH 229T** I can use my knowledge of the co-ordinate system to plot and describe the location of a point on a grid. **MTH 230U**	

SHM Resources					Assessment		Other Resources	Date	Comment
Teaching File page	Textbook	Extension Textbook	Pupil Sheet	Home Activity	Check-Up	Topic Assessment			
340–346	105–108					67			
350–358	109–112		55			14a, b			

Delivering the Curriculum for Excellence

Development planner

Unit	Mathematics 5–14	SHM Topic	Curriculum for Excellence	
PSE Appears throughout and as a separate topic for use from time to time or as sequential pages of work.	**PSE** • **Know that:** – the problem solving and enquiry process can be envisaged as the three broadly interdependent steps of starting a task, doing a task and reporting on a task.	**Problem solving and enquiry** • **Problem solving and enquiry** – uses and applies a range of mathematical knowledge of number properties to solve a variety of problems and puzzles – consolidates and develops the problem solving strategies of 'trial and improvement' and 'listing' – consolidates the use of problem solving strategies in problems and puzzles where selection of an appropriate strategy or strategies is required.		
Information handling 1 and 2	**C/E** • **By selecting sources of information** for tasks, including: – practical experiments – surveys using questionnaires – sampling using a simple strategy. **O/E** • **By designing and using diagrams and tables.** • **By designing or using a database or spreadsheet** with fields defined by pupils with the aid, where appropriate, of a computer package. **D/E** • **By constructing straight line and curved graphs for continuous data** where there is a relationship such as direct proportion – travel, temperature, growth graphs. • **By constructing pie charts of the data** expressed in percentages. With the aid, where appropriate, of a computer package. **I/E** • **From an extended range of displays** (diagrams, tables, graphs, pie charts) **and databases**, retrieving information subject to more than one condition with the aid, where appropriate, of a computer package, involving the use of logical operators (and; or; not) • **By describing the main features of a graph** so as to show an awareness of the significance of the information • **By calculating the average (mean)** to compare sets of data.	**Data handling** • **Surveys and databases** – introduces the importance of designing an efficient questionnaire before using it to collect data in a survey – develops extracting information from a database to include data with up to five conditions. • **Interpreting graphs/charts** – introduce simple pie charts, with 10/12/20/24 or 100 divisions, comparing their effectiveness with that of a compound bar chart – consolidates interpreting and constructing bar charts with class intervals – consolidates interpreting and constructing continuous data straight-line and curved-line graphs – consolidates the range, mode, median and mean of a set of data and develops this to finding the median of an even number of values. • **Probability** – revises language associated with the probability of an event occurring, including: – likelihood: impossible, unlikely, likely certain – chance: no chance, poor chance, even chance, good chance, (and certain) – introduces listing all the possible outcomes of an event, such as rolling a die – introduces the use of language such as 'one in six' to describe the probability of an event happening – considers, in extension activities: – the difference between the theory of outcomes and actual experimental results the meaning of fair, unfair and bias.	Having discussed the variety of ways and range of media used to present data, I can interpret and draw conclusions from the information displayed, recognising that the presentation may be misleading. **MNU 232W** I have carried out investigations and surveys, devising and using a variety of methods to gather information and have worked with others to collate, organise and communicate the results in an appropriate way. **MNU 233W** I can display data in a clear way using a suitable scale, by choosing appropriately from an extended range of tables, charts, diagrams and graphs, making effective use of technology. **MTH 234X** I can use appropriate vocabulary to describe the likelihood of events occurring, using the knowledge and experiences of myself and others to guide me. **MNU 127Y** I can conduct simple experiments involving chance and communicate my predictions and findings using the vocabulary of probability. **MNU 235Y**	
Number 1	**RTN/E** **RN/E**	**Place value** • **Numbers to hundreds of millions** – develops the number sequence to hundreds of millions (8-/9-digit numbers) – includes finding, in relation to ascending and descending sequences, multiples of 10 000, 100 000, 1 000 000, 10 000 000 – develops place value to numbers with up to nine digits – includes adding and subtracting mentally 10/100/1000/10 000/100 000/1 000 000/ 10 000 000/100 000 000 to/from numbers with up to nine digits – deals with: – identifying the large/smaller number in a pair and the largest/smallest number in a set of three 9-digit numbers – ordering up to four non-consecutive 9-digit numbers – finding the number halfway between a pair of multiples of 100 000 or 1 000 000 – reading and writing numbers with up to nine digits – consolidates rounding numbers with up to six digits to the nearest 1000/100	I can use my knowledge of rounding to routinely estimate the answer to a problem, then after calculating, decide if my answer is reasonable, sharing my solution with others. **MNU 201A** I have extended the range of whole numbers I can work with and having explored how decimal fractions are constructed, can explain the link between a digit, its place and its value. **MNU 202B** Having determined which calculations are needed, I can solve problems involving whole numbers using a range of methods, sharing my approaches and solutions with others. **MNU 203C** Having discussed the variety of ways and range of media used to present data, I can interpret and draw conclusions from the information displayed, recognising that the presentation may be misleading. **MNU 232W**	

SHM Resources					Assessment		Other Resources	Date	Comment
Teaching File page	Textbook	Extension Textbook	Pupil Sheet	Home Activity	Check-Up	Topic Assessment			
258–265	69–73								
394–397	114–116		37						
398–406	117–121		38–39			17a, b			
409–414	122–123	20–21	40						
42–55	1–6	1	1–6	1–2		1a, b			

Delivering the Curriculum for Excellence

Development planner

Unit	Mathematics 5–14	SHM Topic	Curriculum for Excellence	
Number 1 (cont.)		– consolidates rounding 8-digit numbers to the nearest million and develops this to include rounding 9-digit numbers to the nearest million – consolidates estimating in multiples of 1000 and introduces estimating in millions, both in the context of information displayed in bar charts.		
Number 2	**AS/E** • **Add:** – mentally for 2-digit numbers including decimals – without a calculator for four digits with at most two decimal places – with a calculator for any number of digits with at most three decimal places in applications in number, measurement and money.	**Addition** • **Mental addition of numbers with up to five digits** ‑ consolidates mental strategies for addition of a pair of: – 2-digit numbers – 3-digit multiples of 10 – 4-digit multiples of a 100 – revises mental addition of a 2-digit number and a 3-digit number – revises mental addition of 3-digit numbers: – bridging a multiple of 10 only – bridging a multiple of 100 only – bridging 1000 only and introduces addition with bridging of both a multiple of 10 and a multiple of 100 – consolidates finding an approximate total based on rounding to the nearest 100/10 and develops this to finding an approximate total based on rounding to the nearest 1000 – introduces finding approximate money totals by: – rounding amounts to the nearest £1/50p – combining two (or more) amounts to make about one or more pounds. • **Mental/written addition involving numbers with four or more digits** – consolidates mental addition of 4-digit numbers: – with no bridging – bridging a multiple of 10 only – consolidates a standard written method of addition involving numbers with up to four digits and develops this to include written addition of a 4-digit number and a 5-digit number – develops this standard written method further, in an extension activity, to include additions involving: – 5-digit numbers – numbers with different numbers of digits – includes the use of a calculator in addition problems involving numbers with six, seven or eight digits.	I can use my knowledge of rounding to routinely estimate the answer to a problem, then after calculating, decide if my answer is reasonable, sharing my solution with others. **MNU 201A** I have extended the range of whole numbers I can work with and having explored how decimal fractions are constructed, can explain the link between a digit, its place and its value. **MNU 202B** Having determined which calculations are needed, I can solve problems involving whole numbers using a range of methods, sharing my approaches and solutions with others. **MNU 203C**	
Number 3	**AS/E** • **Subtract:** – mentally for 2-digit whole numbers including decimals – without a calculator for 4 digits with at most two decimal places – with a calculator for any number of digits with at most three decimal places in applications in number, measurement and money.	**Subtraction** • **Mental subtraction involving numbers with up to five digits** – consolidates mental strategies for subtraction involving certain 4-digit numbers and includes finding the differences between: – a 3-/4-digit multiple of 100 and a 4-digit multiple of 100 – a 3-digit multiple of 100 and any 4-digit number – a 4-digit multiple of 50 and a 4-digit multiple of 100, without bridging a multiple of 1000 – a 3-/4-digit number and a multiple of 1000 – numbers near to and on either side of the same/a different multiple of 1000 – revises mental subtraction of a 2-digit number from a 3-digit number – revises mental subtraction of a 3-digit number: – bridging a multiple of 10 and a multiple of 100 – bridging a multiple of 10 only – and introduces subtraction with bridging of a multiple of 100 only – introduces mental calculation of an approximate difference between two numbers with four or five digits, using rounding to the nearest 1000 and to the nearest 100. • **Mental/written subtraction involving numbers with four or more digits** – consolidates mental subtraction of 4-digit numbers: – with no bridging – bridging a multiple of 10 only – consolidates a standard written method of subtraction involving numbers with up to four digits and develops this to include written subtraction form a 5-digit number of a number with up to four digits – further develops the standard written method, in an extension activity, to include subtractions involving: – 5-digit numbers – different numbers of digits	I can use my knowledge of rounding to routinely estimate the answer to a problem, then after calculating, decide if my answer is reasonable, sharing my solution with others. **MNU 201A** I have extended the range of whole numbers I can work with and having explored how decimal fractions are constructed, can explain the link between a digit, its place and its value. **MNU 202B** Having determined which calculations are needed, I can solve problems involving whole numbers using a range of methods, sharing my approaches and solutions with others. **MNU 203C**	

SHM Resources					Assessment		Other Resources	Date	Comment
Teaching File page	Textbook	Extension Textbook	Pupil Sheet	Home Activity	Check-Up	Topic Assessment			
62–69	7–10		7	3	1				
70–73	11–12	2	8			2a, b			
80–87	13–14		9–10	4	2				
88–93	15–8	3	11			3a, b			

Delivering the Curriculum for Excellence

Development planner

Unit	Mathematics 5–14	SHM Topic	Curriculum for Excellence	

Number 3 (cont.)		– uses and explains standard written methods in problems involving both addition to, and subtraction from, a 5-digit number
		– explores a variety of ways of checking answers to addition and subtraction calculations including:
		  – estimating an approximate answer using rounded numbers
		  – using the inverse operation
		  – doing the calculation in a different order
		  – doing the calculation again in a different way
		– includes the use of a calculator in addition and subtraction problems involving number with six, seven or eight digits.		

Number 4	**MD/E**
	• **Multiply**
	  – mentally for any whole number by a multiple of 10 or 100 (such as 20 or 200)
	  – mentally for any number including decimals by 10, 100, 1000
	  – without a calculator for four digits with at most two decimal places by a single digit
	  – with a calculator for any pair of numbers but at most three decimal places in the answer
	  – in application in number, measurement and money.
	
	FE/E
	• **Solve simple equations and inequations**
		Multiplication
		• **Mental multiplication**
		  – revises mental multiplication of a 2-digit number by a single digit
		  – consolidates and integrates a range of mental multiplication strategies:
		    – involving doubling/halving
		    – based on initial multiplication by 100
		    – involving factors
		  – consolidates and develops mental multiplication by a multiple of 10/100
		  – introduces finding approximate products by rounding before multiplying mentally
		  – develops mental multiplication strategies based on adjusting.
		• **Written multiplication, calculator**
		  – revises multiplication of a 4-digit number by a single digit using an expanded recording
		  – consolidates multiplication of a 4-digit number by a single digit using a shorter standard written method
		  – introduces multiplication of a 3-digit number by a 2-digit number using:
		    – an informal 'cross' method
		    – a standard written method
		  – provides opportunities to use a calculator to solve multiplication problems involving larger numbers
		  – includes, in an extension activity, multiplication of a 4-digit number by a 2-digit number using a standard written method.
			I can use my knowledge of rounding to routinely estimate the answer to a problem, then after calculating, decide if my answer is reasonable, sharing my solution with others.
			MNU 201A
			I have extended the range of whole numbers I can work with and having explored how decimal fractions are constructed, can explain the link between a digit, its place and its value.
			MNU 202B
			Having determined which calculations are needed, I can solve problems involving whole numbers using a range of methods, sharing my approaches and solutions with others.
			MNU 203C
			Having explored the patterns and relationships in multiplication and division, I can investigate and identify the multiples and factors of numbers.
			MTH 207E	

Number 5	**MD/E**
	• **Divide**:
	  – mentally for any whole number by a multiple of 10 or 100 (such as 20 or 200)
	  – mentally for any numbers including decimals by 10, 100, 1000
	  – without a calculator for four digits with at most two decimal places by a single digit
	  – with a calculator for any pair of numbers but at most three decimal places in the answer
	  – in applications in number, measure and money.
		Division
		• **Division by a single digit**
		  – revises and develops mental strategies for division by a single digit of certain 4-digit numbers and numbers just beyond the extent of multiplication tables
		  – consolidates and develops written division by a single digit of 4-/5-digit numbers, using a short standard method
		  – provides division problems that also include the use of other operations
		• **Division by two digits**
		  – deals with mental division of a 3-digit multiple of 10 by a 2-digit multiple of 10
		  – includes rounding:
		    – of a 3-digit dividend to the nearest 100
		    – of a 2-digit divisor to the nearest 10
		  to facilitate mental division
		  – introduces division, exact and with remainders, of a 3-digit number by a 2-digit number using a standard written method (2-digit quotients)
		  – deals with methods for checking the answers to written division calculations
		  – provides opportunities for the children to use a calculator to solve word problems involving large numbers, where division is used in combination with other operations
		  – includes an extension activity dealing with division of a 4-digit number by a 2-digit number, using a standard written method.
			I can use my knowledge of rounding to routinely estimate the answer to a problem, then after calculating, decide if my answer is reasonable, sharing my solution with others.
			MNU 201A
			I have extended the range of whole numbers I can work with and having explored how decimal fractions are constructed, can explain the link between a digit, its place and its value.
			MNU 202B
			Having determined which calculations are needed, I can solve problems involving whole numbers using a range of methods, sharing my approaches and solutions with others.
			MNU 203C
			Having explored the patterns and relationships in multiplication and division, I can investigate and identify the multiples and factors of numbers.
			MTH 207E	

Number 6	**PS/E**
	• **Continue and describe sequences:**
	  – involving square and triangular numbers
	  – find specified items in sequences
	  – prime numbers.
		Number properties
		• **Number types, sequences, patterns**
		  – revises language and notation associated with square numbers
		  – consolidates the idea of a smallest/lowest common multiple for a pair of numbers
		  – consolidates continuing number sequences, and using 'rules' to describe or generate number sequences, including sequences which increase or decrease:
		    – in equal steps of 29/31/39/41/49/51 and 'teens'
		    – in other ways
			Having explored more complex number sequences, including well-known named number patterns, I can explain the rule used to generate the sequence, and apply it to extend the pattern.
			MTH 221P
			I can continue and devise more involved repeating patterns or designs, using a variety of media.
			MTH 115P
			Through exploring number patterns, I can recognise and continue simple number sequences and can explain the rule I've applied.
			MTH 116P	

SHM Resources					Assessment		Other Resources	Date	Comment
Teaching File page	Textbook	Extension Textbook	Pupil Sheet	Home Activity	Check-Up	Topic Assessment			
100–107	19–21			5					
108–113	22–24	4, 22	12, 13			4a, b			
122–125	25–27			6					
126–132	28–30	5	14–16			5a, b, c			
140–151	31–35		17	7–8					

Delivering the Curriculum for Excellence

Development planner

Unit	Mathematics 5–14	SHM Topic	Curriculum for Excellence	
Number 6 (cont.)	**RTN/E** • **Work with:** - negative numbers (e.g. temperatures) **AS/E** • **Add and subtract:** - positive and negative numbers in applications such as rise in temperature. **FW/E** • **Use a 'function machine' to reverse for inverse operations.** • **Use notation** to describe general relationships between two sets of numbers. • **Use and devise simple rules.**	– introduces the triangular numbers 1, 3, 6, 10, 15, 21…105, 120 – consolidation ordering and addition and subtraction of negative numbers, in the context of temperature – revises listing factors and factor pairs – explores prime numbers to 50 and then to 100 and includes expressing any number to 100 as a product of its prime factors – revises methods (including consideration of the last digit-sum) of testing numbers, without dividing, for exact divisibility by 2, 3, 4, 5, 6, 8, 9, 10, 25 and 100 – introduces, for larger numbers, an alternative test for divisibility by 8 and an extension of the test for divisibility by 9. • **Formulae in words and symbols** – revises finding the 'rule', or word formula, to describe a relationship between two sets of numbers – introduces using letters to express a rule as symbolic formula – introduces using a formula with two variables to calculate the value of one variable, given the value of the other.	Having explored more complex number sequences, including well-known named number patterns, I can explain the rule used to generate the sequence, and apply it to extend the pattern. **MTH 221P** I can continue and devise more involved repeating patterns or designs, using a variety of media. **MTH 115P** Through exploring number patterns, I can recognise and continue simple number sequences and can explain the rule I've applied. **MTH 116P** I can show my understanding of how the number line extends to include numbers less than zero and have investigated how these numbers occur and are used. **MNU 206D** Having determined which calculations are needed, I can solve problems involving whole numbers using a range of methods, sharing my approaches and solutions with others. **MNU 203C**	
Number 7	**RTN/E** • **Work with:** – all widely used fractions and equivalences among these and decimals (in applications). **FPR/E** • **Work with fractions:** – mentally find widely-used fractions of whole number quantities – with a calculator find a fraction of a quantity – without a calculator as previously defined • **Find ratios between quantities.** • **Use simple unitary ratio.**	**Fractions** • **Equivalence, thousandths** – consolidates forming equivalent fractions by multiplying numerator and denominator – develops simplifying fractions by dividing numerator and denominator to include fractions where more than one division may be necessary – consolidates converting between mixed number and improper fractions – introduces thousandths, expressing several mm, m, g and ml as fractions of 1 m, 1 km, 1 kg, and 1 ℓ respectively, and including mixed numbers. • **Addition, subtraction and multiplication** – introduces addition involving: – proper fractions (same denominator) that total less than one – a mixed number and a proper fraction (same denominator) where the fractional parts total less than one/more than one – introduces subtraction involving: – proper fractions (same denominator) – a mixed number and a proper fraction (same denominator) in which the fractional part of the mixed number is smaller/greater than the fraction to be subtracted. – introduces addition and subtraction involving halves and quarters – includes, in an extension activity, addition and subtraction involving fractions and mixed numbers that have **different** denominators – consolidates finding 'any' fraction, numerator ≥ 1, of a set/quantity – consolidates expressing one quantity as a fraction of another – introduces multiplication involving a whole number times a fraction. • **Ratio and proportion** – introduces comparing 'part to part' and 'part to whole' and associated language – introduces the term ratio and includes expressing a given ratio in simplest form – introduces, as extension, the term proportion.	I can use my knowledge of rounding to routinely estimate the answer to a problem, then after calculating, decide if my answer is reasonable, sharing my solution with others. **MNU 201A** I have extended the range of whole numbers I can work with and having explored how decimal fractions are constructed, can explain the link between a digit, its place and its value. **MNU 202B** Having determined which calculations are needed, I can solve problems involving whole numbers using a range of methods, sharing my approaches and solutions with others. **MNU 203C** I have explored the contexts in which problems involving decimal fractions occur and can solve related problems using a variety of methods. **MNU 204C** Having explored the patterns and relationships in multiplication and division, I can investigate and identify the multiples and factors of numbers. **MTH 207E** I have investigated the everyday contexts in which simple fractions, percentages or decimal fractions are used and can carry out the necessary calculations to solve related problems. **MNU 208H** I can show the equivalent forms of simple fractions, decimal fractions and percentages and can choose my preferred form when solving a problem, explaining my choice of method. **MNU 209** I have investigated how a set of equivalent fractions can be created, understanding the meaning of simplest form, and can apply my knowledge to compare and order the most commonly used fractions. **MTH 210H** I can show how quantities that are related can be increased or decreased proportionally and apply this to solve problems in everyday contexts. **MNU 311J**	
Number 8	**RTN/E** • **Work with:** – all widely used fractions and equivalences among these and decimals (in applications); – decimals to three places (practical applications in measurement). **RN/E** • **Round any number** to one decimal place. **AS/E** • **Add and subtract:** – mentally for 2-digit numbers including decimals – without a calculator for four digits with at most two decimal places – with a calculator for any number of digits with at most three decimal places – in applications in number, measurement and money.	**Decimals** • **1- and 2-place decimals: addition, subtraction and multiplication** – revises sequences, comparing and ordering for 1- and 2-place decimals – consolidates and develops mental strategies for the addition and subtraction 1-/2-place decimals – revises rounding a 1-place decimal to the nearest whole number and introduces rounding a 2-place decimal: – to the nearest tenth/first decimal place – to the nearest whole number – includes calculating mentally an approximate sum/difference by rounding to the nearest whole number – consolidates and develops a standard written method for addition and subtraction of 1-/2-place decimals – revises mental multiplication of 1-/2-place decimals by a single digit, 10, 100 or a multiple of 10 and introduces mental multiplication by a multiple of 100 and by 1000 – consolidates and develops a standard written method for multiplication of larger number with one/two decimals places by a single digit.	I can use my knowledge of rounding to routinely estimate the answer to a problem, then after calculating, decide if my answer is reasonable, sharing my solution with others. **MNU 201A** I have extended the range of whole numbers I can work with and having explored how decimal fractions are constructed, can explain the link between a digit, its place and its value. **MNU 202B** Having determined which calculations are needed, I can solve problems involving whole numbers using a range of methods, sharing my approaches and solutions with others. **MNU 203C** I have explored the contexts in which problems involving decimal fractions occur and can solve related problems using a variety of methods. **MNU 204C** Having explored the patterns and relationships in multiplication and division, I can investigate and identify the multiples and factors of numbers. **MTH 207E** I have investigated the everyday contexts in which simple fractions, percentages or decimal fractions are used and can carry out the necessary calculations to solve related problems. **MNU 208H** I can show the equivalent forms of simple fractions, decimal fractions and percentages and can choose my preferred form when solving a problem, explaining my choice of method. **MNU 209H**	

Teaching File page	SHM Resources		Pupil Sheet	Home Activity	Assessment		Other Resources	Date	Comment
	Textbook	Extension Textbook			Check-Up	Topic Assessment			
152–156	36–37					6a, b			
166–173	38–39		18	9	3				
174–184	40–44	6			4				
185–188	45–46	7				7a, b			
198–208	47–52		19–21	10–11	5–6				

Delivering the Curriculum for Excellence

Development planner

Unit	Mathematics 5–14	SHM Topic	Curriculum for Excellence	

Unit	Mathematics 5–14	SHM Topic	Curriculum for Excellence
Number 8 (cont.)	**MD/E** • **Multiple and divide:** – mentally for any numbers including decimals by 10, 100 or 1000 – without a calculator for 4 digits with at most 2 decimal places by a single digit – with a calculator for any pair of numbers but at most 3 decimal places in the answer – in applications in number, measurement and money.	• **1- and 2-place decimals: division** – revises mental division, leading to a quotient with one decimal place, of: – a 1-/2-digit whole number by 10 – a 2-digit number with one decimal place by 2–9 – introduces mental division, leading to a quotient with two decimal places, of: – a 2-digit number with one decimal place by 10 – a 2-digit whole number by 100 – a 3-digit number with two decimal places by 2–9 – develops, in context, the ability to round money amounts to the nearest pound to calculate mentally approximate products and quotients – revises the use of a standard written method for division by 2–9 of a 3-digit number with one decimal place and develops this to include 4-digit numbers – introduce the use of a standard written method for division by 2–9 of a 3-/4-digit number with two decimal places – involves the use of standard written methods of multiplication and division in word problems where the appropriate operation must be selected – includes using and applying calculator skills in word problems, where the appropriate operation/s must be selected, and in division which involve: – rounding a number with several decimal places to the nearest whole number and, depending on the context, to the nearest appropriate whole number – calculating an exact remainder from a quotient with several decimal places.	I can use my knowledge of rounding to routinely estimate the answer to a problem, then after calculating, decide if my answer is reasonable, sharing my solution with others. **MNU 201A** I have extended the range of whole numbers I can work with and having explored how decimal fractions are constructed, can explain the link between a digit, its place and its value. **MNU 202B** Having determined which calculations are needed, I can solve problems involving whole numbers using a range of methods, sharing my approaches and solutions with others. **MNU 203C** I have explored the contexts in which problems involving decimal fractions occur and can solve related problems using a variety of methods. **MNU 204C** Having explored the need for rules for the order of operations in number calculations, I can apply them correctly when solving simple problems. **MTH 205C** Having explored the patterns and relationships in multiplication and division, I can investigate and identify the multiples and factors of numbers. **MTH 207E** I have investigated the everyday contexts in which simple fractions, percentages or decimal fractions are used and can carry out the necessary calculations to solve related problems. **MNU 208H** I can show the equivalent forms of simple fractions, decimal fractions and percentages and can choose my preferred form when solving a problem, explaining my choice of method. **MNU 209H**
Number 9	**RTN/E** • **Work with:** – all widely used fractions and equivalence among these and decimals (in applications); – decimals to three places (practical applications in measurement). **RN/E** • **Round any number** to one decimal place. **AS/E** • **Add and subtract:** – mentally for 2-digit numbers including decimals – without a calculator for four digits with at most two decimal places – with a calculator for any number of digits with at most three decimal places – in application in number, measurement and money. **MD/E** • **Multiply and divide:** – mentally for any numbers including decimals by 10, 100 or 1000 – without a calculator for four digits with at most two decimal places by a single digit – with a calculator for any pair of numbers but at most three decimal places in the answer – in applications in number, measurement and money.	• **3-place decimals** – introduces decimal notation for thousandths in 4- and 5-digit numbers, and includes work on sequences – links the decimal notation to the corresponding notation for fractions and mixed numbers in the context of measures – develops place value to thousandths and includes work on comparing and ordering 3-place decimals – deals with rounding a 3-place decimal – to the nearest tenth/first decimal place – to the nearest whole number and applies the latter to calculating mentally approximate sums and differences – includes using and applying activities which involve: – changing larger metric units, expressed as 3–place decimals, to smaller units, expressed as whole numbers, and vice-versa – changing a fraction to decimal form, by dividing the numerator by the denominator, and using this to compare two fractions – uses, in extension activities, standard written methods for adding and subtracting 3-place decimals.	I can use my knowledge of rounding to routinely estimate the answer to a problem, then after calculating, decide if my answer is reasonable, sharing my solution with others. **MNU 201A** I have extended the range of whole numbers I can work with and having explored how decimal fractions are constructed, can explain the link between a digit, its place and its value. **MNU 202B** Having determined which calculations are needed, I can solve problems involving whole numbers using a range of methods, sharing my approaches and solutions with others. **MNU 203C** I have explored the contexts in which problems involving decimal fractions occur and can solve related problems using a variety of methods. **MNU 204C** Having explored the need for rules for the order of operations in number calculations, I can apply them correctly when solving simple problems. **MTH 205C** Having explored the patterns and relationships in multiplication and division, I can investigate and identify the multiples and factors of numbers. **MTH 207E** I have investigated the everyday contexts in which simple fractions, percentages or decimal fractions are used and can carry out the necessary calculations to solve related problems. **MNU 208H** I can show the equivalent forms of simple fractions, decimal fractions and percentages and can choose my preferred form when solving a problem, explaining my choice of method. **MNU 209H**

SHM Resources					Assessment		Other Resources	Date	Comment
Teaching File page	Textbook	Extension Textbook	Pupil Sheet	Home Activity	Check-Up	Topic Assessment			
209–217	53–59			12	7–8				
218–229	60–63	8–9	22–23, 20	13	9	8a, b, c			

Delivering the Curriculum for Excellence

Development planner

Unit	Mathematics 5–14	SHM Topic	Curriculum for Excellence	

Number 10	RTN/E
	FPR/E
	• **Work with percentages:**
	  – mentally find widely-used percentages of whole number quantities
	  – with a calculator find a percentage of a quantity
	  – without a calculator as previously defined.
		Percentages
		• **Percentages**
		  – revises the concept, language and notation associated with percentages
		  – consolidates and extends the links between the notation for percentages and the corresponding notation for fractions and decimals
		  – deals with finding a percentage of a set/quantity using knowledge of equivalent fractions
		  – discusses strategies for finding, for example:
		  – 5% (by first finding 10% and then halving)
		  – 12½% (by first finding 25% and then halving)
		  – 14% (by finding, and then adding, 10% and 4%)
		  – introduces comparing and ordering numbers expressed as percentages, fractions or decimals
		  – provides opportunities for the children to use and apply their knowledge about percentages to solve word problems in a range of contexts
		  – introduces finding a percentage of a set/quantity using a calculator, rounding the 1-/2-place decimals in the display to the nearest whole number
		  – introduces, in extension activities:
		    – finding a percentage greater than 100% of a set/quantity
		    – expressing more difficult fractions as percentages using a calculator and rounding the numbers (with several decimal places) in the display to the nearest 1%.
			I can use my knowledge of rounding to routinely estimate the answer to a problem, then after calculating, decide if my answer is reasonable, sharing my solution with others.
			MNU 201A
			I have extended the range of whole numbers I can work with and having explored how decimal fractions are constructed, can explain the link between a digit, its place and its value.
			MNU 202B
			Having determined which calculations are needed, I can solve problems involving whole numbers using a range of methods, sharing my approaches and solutions with others.
			MNU 203C
			I have explored the contexts in which problems involving decimal fractions occur and can solve related problems using a variety of methods.
			MNU 204C
			Having explored the need for rules for the order of operations in number calculations, I can apply them correctly when solving simple problems.
			MTH 205C
			Having explored the patterns and relationships in multiplication and division, I can investigate and identify the multiples and factors of numbers.
			MTH 207E
			I have investigated the everyday contexts in which simple fractions, percentages or decimal fractions are used and can carry out the necessary calculations to solve related problems.
			MNU 208H
			I can show the equivalent forms of simple fractions, decimal fractions and percentages and can choose my preferred form when solving a problem, explaining my choice of method.
			MNU 209H	

Measure 2	RTN/E
	• **Work with:**
	  – decimals to three places (practical applications in measurement).
	ME/E
	• **Measure using standard units:**
	  – accuracy and device as appropriate to the application.
	• **Estimate measurements.**
	• **Work with tonne** when appropriate.
	• **Read scales** on measuring devices, including estimating between graduations.
		Measure
		• **Weight**
		  – revises reading and recording weights to the nearest graduation mark on scales showing tenths of a kilogram in decimal form and in multiples of 100 g
		  – introduces reading scales by estimating between marks showing:
		    – non-consecutive numbers of grams or kilograms
		    – consecutive numbers of kilograms (expressing the reading to the nearest $\frac{1}{10}$ kg or 0.1 kg or 100 g
		    – consecutive numbers of tenths of kilograms (expressing the reading to the nearest $\frac{1}{100}$ kg or 0.01 kg or 10 g)
		  – introduces the relationship 1 tonne = 1000 kilograms and includes recording weights given in tonnes and kilograms in kilograms and vice-versa
		  – provides activities which encourage the children to use and apply their knowledge of weight and includes problem solving tasks that involve practical estimating and weighing and choosing the most suitable weighing apparatus.
			I can use my knowledge of the sizes of familiar objects or places to assist me when making an estimate of measure.
			MNU 217M
			I can use the common units of measure, convert between related units of the metric system and carry out calculations when solving problems.
			MNU 218M	

Measure 1	RTN/E
	• **Work with:**
	  – decimals to three places (practical applications in measurement).
	ME/E
	• **Measure and draw using standard units:**
	  – accuracy and device as appropriate to the application.
	• **Estimate measurements:**
	  – small lengths in millimetres
	  – larger lengths in metres.
	• **Read scales** on measuring devices, including estimating between graduations.
	PFS/E
	• **Calculate using rules.**
	• **Use scales** such as 1 cm to 1, 2, 5 or 10 m: or represented by a ratio such as 1:100 to interpret or draw maps, plans, diagrams, or to make models.
		Measure
		• **Length:**
		  – consolidates estimating and measuring lengths in metres and centimetres, including recording in different ways
		  – introduces scale drawing using scales
		  – involves using practical skills to choose appropriate units and measuring devices and applying knowledge to solve problems that include using the relationships 1000 m = 1 km and 10 mm = 1 cm
		  – revises finding the perimeters of shapes both by measuring to the nearest cm/½cm/mm and by calculating, and consolidates the use of a word formula for finding the perimeter of a rectangle.
			I can use my knowledge of the sizes of familiar objects or places to assist me when making an estimate of measure.
			MNU 217M
			I can use the common units of measure, convert between related units of the metric system and carry out calculations when solving problems.
			MNU 218M
			I can explain how different methods can be used to find the perimeter and area of a simple 2D shape or volume of a simple 3D object.
			MNU 219M	

SHM Resources					Assessment		Other Resources	Date	Comment
Teaching File page	Textbook	Extension Textbook	Pupil Sheet	Home Activity	Check-Up	Topic Assessment			
240–249	64–68	10–11	24	14		9a, b			
272–279	74–76		25–27	15		10			
284–293	77–82		28			11			

Delivering the Curriculum for Excellence

Development planner

Unit	Mathematics 5–14	SHM Topic	Curriculum for Excellence	
Measure 3	T/E	**Time** • **Time: 24-hour times** – revises relationships among units of time from a millennium to a minute and links units of time with typical periods they are used to measure – consolidates reading and writing times in 1 minute intervals using 24-hour notation and includes covering between 12-hour and 24-hour times – introduces World time zones and the concept of local time as a context for using and applying skills in reading and writing times using 24-hour notation, with the possibility for some children, of briefly introducing *British Summer Time*, *Greenwich Mean Time* and the *International Date Line*.	I can tell the time using 12 and 24 hour clocks, explain how it impacts on my daily routine and ensure that I am organised and ready for events throughout my day. **MNU 109L** I can use and interpret electronic and paper-based timetables and schedules to plan events and activities, and make time calculations as part of my planning **MNU 214L**	
Measure 5	T/E • **Time activities** with a digital stopwatch in seconds, tenths, hundredths.	**Time: durations, rate, speed** – consolidates finding times in multiples of 5 minutes before/after times displayed in 24–hour notation: – bridging one hour – bridging more than one hour – reinforces finding durations in multiples of 5 minutes between digital times given in 24-hour notation, bridging one hour/more than one hour and, in an extension activity, develops this to include finding durations in multiples of 1 minute – includes problems that require the children to use and apply the above in a variety of contexts – introduces the concept of rate and speed with rate being expressed in actions per second, speed in metres per second, miles per hour and, in a suggested follow-up activity, in kilometres per hour.	I can use and interpret electronic and paper-based timetables and schedules to plan events and activities, and make time calculations as part of my planning **MNU 214L** I can carry out practical tasks and investigations involving timed events and can explain which unit of time would be most appropriate to use. **MNU 215L** Using simple time periods, I can give a good estimate of how long a journey should take, based on my knowledge of the link between time, speed and distance. **MNU 216L**	
Measure 6	RTN/E • **Work with:** – decimals to three places (practical applications in measurement). ME/E • **Measuring using standard units:** – accuracy and device as appropriate to the application. • **Read scales** on measuring devices including estimating between graduations. PFS/E • **Calculating using rules:** – volumes of cuboids and cubes.	**Measure** • **Volume/Capacity** – introduces the centilitre and the relationships 1 ℓ = 100 cl, 1 cl = 10 ml – includes the use of decimal notation for the tenth of a litre and the relationships 100 ml = $\frac{1}{10}$ ℓ = 0.1 ℓ, 500 ml = $\frac{1}{2}$ ℓ = 0.5 ℓ … – revises reading scales marked in 100 ml, 50 ml, 20 ml and 10 ml divisions – provides practical estimating and measuring activities using centilitres – includes problems which require the children to use and apply the knowledge of metric units to volume – introduces finding the volume, in cubic centimetres, or a cuboid, by initially counting cubes, then by the application of the formula V = l × b × h. • **Mixed measure** – consolidates and develops knowledge about commonly used imperial units and their metric equivalents: – revises miles, pints and gallons – introduces inches, feet and yards and pounds and ounces.	I can use my knowledge of the sizes of familiar objects or places to assist me when making an estimate of measure. **MNU 217M** I can use the common units of measure, convert between related units of the metric system and carry out calculations when solving problems. **MNU 218M** I can explain how different methods can be used to find the perimeter and area of a simple 2D shape or volume of a simple 3D object. **MNU 219M**	
Measure 4	ME/E • **Estimate measurements:** – areas in square metres. • **Work with square kilometre, hectare,** when appropriate. PFS/E • **Calculate using rules:** – areas of rectangles and squares.	**Measure** • **Area** – revises using the letter formula A = l × b for finding the area of a rectangle – introduces the hectare and the square kilometre and the relationships 1 hectare = 10 000 m², 1 square kilometre = 1 000 000 square metres and 1 square kilometre = 100 hectares – consolidates finding the area of a right-angled triangle drawn on a squared grid and introduces a method of calculating the area when the triangle is not drawn on a squared grid – includes finding the area, in cm² and m², of a composite shape by sub-dividing it into rectangles, squares and right-angle triangles – provides extension activities which lead to finding the area of an isosceles triangle using the formula A = $\frac{1}{2}$ × b × h.	I can use my knowledge of the sizes of familiar objects or places to assist me when making an estimate of measure. **MNU 217M** I can use the common units of measure, convert between related units of the metric system and carry out calculations when solving problems. **MNU 218M** I can explain how different methods can be used to find the perimeter and area of a simple 2D shape or volume of a simple 3D object. **MNU 219M**	
Shape 3	RS/E • **Use properties of 3D shapes:** – make 3D models, solid or skeletal, including using nets: triangular prism, pyramid, tetrahedron.	**Shape** • **3D shape** – introduces a relationship involving the number of faces, vertices and edges of a polyhedra – develops recognising 3D shapes from their nets – introduces visualising 3D shapes in different ways (plan, front and side views) – introduces the use of 'perpendicular' and 'parallel' to describe properties of edges of 3D shapes – includes, in an extension activity, a practical investigation involving building and fitting 3D shapes.	Having explored a range of 3D objects and 2D shapes, I can use mathematical language to describe their properties, and through investigation can discuss where and why particular shapes are used in the environment. **MTH 223S** Through practical activities, I can show my understanding of the relationship between 3D objects and their nets. **MTH 224S** I can draw 2D shapes and make representations of 3D objects using an appropriate range of methods and efficient use of resources. **MTH 225S**	

SHM Resources					Assessment		Other Resources	Date	Comment
Teaching File page	Textbook	Extension Textbook	Pupil Sheet	Home Activity	Check-Up	Topic Assessment			
298–303	83–84		29	16					
304–311	85–90	12–13		17		12a, b			
318–323	91–94					13			
325–328	95–96								
332–339	97–99	14	30			14			
346–352	100–102	15							

Delivering the Curriculum for Excellence

Development planner

Unit	Mathematics 5–14	SHM Topic	Curriculum for Excellence	
Shape 1	**S/E** • **Work with symmetry:** – determine whether or not shapes have rotational symmetry.	**Shape** • **2D shape: rotational symmetry** – revises reflecting straight lines to create patterns with horizontal and vertical lines of symmetry – introduces rotational symmetry involving simple shapes and regular polygons – introduces completing and creating coloured patterns which have rotational symmetry.	I can illustrate the lines of symmetry for a range of 2D shapes and apply my understanding to create and complete symmetrical pictures and patterns. **MTH 231V** I can continue and devise more involved repeating patterns or designs, using a variety of media. **MTH 115P** Having investigated patterns in the environment, I can use appropriate mathematical vocabulary to discuss the rotational properties of shapes, pictures and patterns and can apply my understanding when completing or creating designs. **MTH 431V**	
Shape 2	**RS/E** • **Use properties of 2D shapes:** – discuss the side, angle, diagonal properties of quadrilaterals: square, rectangle, rhombus, parallelogram, kite, trapezium – define and classify quadrilaterals – relate diameter and circumference (practical work only). • **Draw triangles:** – given three sides, two sides and included angle, two angles and one side.	**Shape** • **2D shape: properties, patterns, construction** – consolidates naming and describing quadrilaterals using properties of angles, sides and symmetry – introduces the diagonal properties of quadrilaterals along with associated terms 'bisect' and 'intersect' – deals with identifying triangles and quadrilaterals given information about the side, angle, diagonal and/or symmetry properties – includes, in extension activities, drawing tile patterns based on two different shapes and using ruler and compasses to construct isosceles, equilateral and scalene triangles.	Having explored a range of 3D objects and 2D shapes, I can use mathematical language to describe their properties, and through investigation can discuss where and why particular shapes are used in the environment. **MTH 223S** I can draw 2D shapes and make representations of 3D objects using an appropriate range of methods and efficient use of resources. **MTH 225S** I have investigated angles in the environment, and can discuss, describe and classify angles using appropriate mathematical vocabulary. **MTH 226T** I can accurately measure and draw angles using appropriate equipment, applying my skills to problems in context. **MTH 227T** I can name angles and find their sizes using my knowledge of the properties of a range of 2D shapes and the angle properties associated with intersecting and parallel lines. **MTH 323T**	
Shape 4	**PM/E** • **Discuss position and movement:** – use bearings and distances to produce accurate scale drawings of routes – use co-ordinates in all four quadrants to plot position – calculate distances along grid lines. **S/E** • **Work with symmetry:** – move a tile of a shape on a squared grid in order to translate, reflect or rotate the shape. **A/E** • **Angle:** – use 'reflex' to describe angles – know the sum of the angles of a triangle is two right angles. **PFS/E** • **Use scales** such as 1 cm to 1, 2, 5 or 10 m to interpret or draw maps, plans, diagrams.	**Shape** • **Position, movement and angle** – revises co-ordinates in the first quadrant and introduces co-ordinates in all four quadrants, and their notation – includes drawing/completing on a four-quadrant co-ordinate grid: – various polygons – shapes which have been reflected in one or two mirror lines (which are also axes of the co-ordinate grid) – shapes which have been translated – shapes which have been rotated about a vertex that lies at the origin – consolidates acute and obtuse angles and introduces reflex angles – deals with estimating angles to the nearest 5° and measuring/drawing angles, with a protractor, to the nearest degree – includes calculating the size of an angle based on the knowledge that: – the sum of the interior angles of a triangle is 180° – the number of degrees in one complete turn is 360° – revises bearings and the associated 3-figure notation and develops accuracy by including measuring bearings to the nearest degree – includes, as extension activities: – reflecting shapes in one or two mirror lines on a 4-quadrant co-ordinate grid, where the lines of symmetry are not also the axes of the co-ordinate grid – measuring bearings on a map to the nearest degree and calculating distances using the scale: 1 cm to 250 m.	I can use my knowledge of the co-ordinate system to plot and describe the location of a point on a grid. **MTH 230U** I can plot and describe the position of a point on a 4–quadrant co-ordinate grid. **MTH 429U** I can apply my understanding of the 4–quadrant co-ordinate system to move, and describe the transformation of, a point or shape on a grid. **MTH 430U** Having investigated patterns in the environment, I can use appropriate mathematical vocabulary to discuss the rotational properties of shapes, pictures and patterns and can apply my understanding when completing or creating designs. **MTH 431V** I have investigated angles in the environment, and can discuss, describe and classify angles using appropriate mathematical vocabulary. **MTH 226T** I can accurately measure and draw angles using appropriate equipment, applying my skills to problems in context. **MTH 227T** I have explored the relationships that exist between the sides, or sides and angles, in right angled triangles and can select and use an appropriate strategy to solve related problems, interpreting my answer for the context. **MTH 425S** Having investigated navigation in the world, I can apply my understanding of bearings and scale to interpret maps and plans and create accurate plans, and scale drawings of routes and journeys. **MTH 324T** Having investigated where, why and how scale is used and expressed, I can apply my understanding to interpret simple models, maps and plans. **MTH 229T** I can name angles and find their sizes using my knowledge of the properties of a range of 2D shapes and the angle properties associated with intersecting and parallel lines. **MTH 323T**	

SHM Resources					Assessment		Other Resources	Date	Comment
Teaching File page	Textbook	Extension Textbook	Pupil Sheet	Home Activity	Check-Up	Topic Assessment			
356–360	103–104		31–33						
361–365	105–106	16–17	34			15			
372–388	107–113	18, 19	35–36			16a, b			

Delivering the Curriculum for Excellence

Block Planners

SHM 5   Year:

Class: **Group:**   **Block 1**   **Date:**

WEEK	ORAL/MENTAL	MATHEMATICS TOPIC
1	**Numbers in 100 thousands**	**Number unit 1:** Numbers to 100 000
2		
3		
4		
		ASSESSMENT
5		**Measure unit 2:** Length
6		**Shape unit 2:** 2D shape: properties and patterns **Shape unit 3:** 2D shape: link symmetry

Class: **Group:**   **Block 2**   **Date:**

WEEK	ORAL/MENTAL	MATHEMATICS TOPIC
1	**Addition in 1000** Revise **Numbers to 10 000**	**Number unit 2:** Addition to 1000
2		
3		
		ASSESSMENT
4		**Number unit 7:** Money
5		**Time unit 1:** Minutes past/to the hour
6		**Information handling unit 1:** Bar charts

SHM 5   Year:

Class: **Group:**   **Block 3**   **Date:**

WEEK	ORAL/MENTAL	MATHEMATICS TOPIC
1	**Subtraction to 1000** Revise **Addition to 1000**	**Number unit 3:** Subtraction to 1000
2		
3		
4		
		ASSESSMENT
5		**Shape unit 1:** 3D shape
6		**Measure unit 1:** Weight

Class: **Group:**   **Block 4**   **Date:**

WEEK	ORAL/MENTAL	MATHEMATICS TOPIC
1	**Multiplication tables facts**	**Number unit 4:** Multiplication
2		
3		
		ASSESSMENT
4		**Number unit 9:** Number properties
5		**Information handling unit 2:** Pictograms
6		**Time unit 3:** Durations

Delivering the Curriculum for Excellence

SHM 5   Year:

Class: **Group:**   **Block 5**   **Date:**

WEEK	ORAL/MENTAL	MATHEMATICS TOPIC
1	**Division facts**	**Number unit 5:** Division
2		
3		
4		**Number unit 6:** Fractions
5		
6		**Measure unit 3:** Capacity

Class: **Group:**   **Block 6**   **Date:**

WEEK	ORAL/MENTAL	MATHEMATICS TOPIC
1	**All sections**	**Number unit 8:** Decimals
2		
3		
4		
		ASSESSMENT
5		**Shape unit 4:** Position, movement and angle
6		**Measure unit 4:** Area

SHM 5 Year:

Class: Group: Block: Date:

WEEK	ORAL/MENTAL	MATHEMATICS TOPIC
1		
2		
3		
4		
5		
6		

Class: Group: Block: Date:

WEEK	ORAL/MENTAL	MATHEMATICS TOPIC
1		
2		
3		
4		
5		
6		

-

Delivering the Curriculum for Excellence

SHM 6   Year:

Class: **Group:**   **Block 1**   **Date:**

WEEK	ORAL/MENTAL	MATHEMATICS TOPIC
1	• **Adding/subtracting powers of 10 to/from numbers with up to eight digits** • **Multiplying/dividing by 10, 100 and 1000**	**Number unit 1:** Numbers to millions
2		
3		
4		**Measure unit 3:** Length
5		**Shape unit 1:** 2D shape: line symmetry
6		**Measure unit 4:** Weight

Class: **Group:**   **Block 2**   **Date:**

WEEK	ORAL/MENTAL	MATHEMATICS TOPIC
1	• **Adding two-/three-digit numbers** • **Adding four-digit multiples of 100**	**Number unit 2:** Addition
2		
3	• **Subtracting two-/three-digit numbers** • **Subtracting from four-digit multiples of 100 and 1000**	**Number unit 3:** Subtraction
4		
5		**Measure unit 1:** Time: reading and writing times
6		**Shape unit 2:** 2D shape: properties, puzzles and patterns

SHM 6   Year:

Class: **Group:**   **Block 3**   **Date:**

WEEK	ORAL/MENTAL	MATHEMATICS TOPIC
1	• **Tables facts** • **Multiplying by multiples of 1, 100 and 1000**	**Number unit 4:** Multiplication
2	• **Multiplying a two-digit number by a single digit** • **Multiplication strategies**	
3		
4		**Measure unit 6:** Area
5		**Data handling unit 1:** Interpreting graphs/data
6		

Class: **Group:**   **Block 4**   **Date:**

WEEK	ORAL/MENTAL	MATHEMATICS TOPIC
1	• **Division facts** • **Halving three-/four-digit numbers** • **Dividing numbers 'beyond the tables' by a single digit**	**Number unit 5:** Division
2		
3		
4	• **Continuing number sequences** • **Using 'rules' to describe or generate number sequences** • **'Testing' numbers for exact divisibility by 2, 3, 4, 5, 6, 8, 9, 10 and 100**	**Number unit 6:** Number properties
5		
6		**Measure unit 2:** Time: durations, seconds **Shape unit 3:** 3D shape

Delivering the Curriculum for Excellence

SHM 6   Year:

Class: **Group:**   **Block 5**   **Date:**

WEEK	ORAL/MENTAL	MATHEMATICS TOPIC
1	• **Converting mixed numbers to improper fractions and vice versa** • **Forming equivalent fractions by multiplying/dividing numerator and denominator**	**Number unit 7:** Fractions
2		
3	• **Adding/subtracting two-digit numbers with one decimal place** • **Multiplying/dividing two digit numbers with one decimal place by a single digit**	**Number unit 6:** Decimals: tenths
4		
5		**Measure unit 5**: Volume/Capacity
6		**Shape unit 4:** Position, movement and angle

Class: **Group:**   **Block 6**   **Date:**

WEEK	ORAL/MENTAL	MATHEMATICS TOPIC
1	• **Adding/subtracting three-digit numbers with two decimal places** • **Multiplying two-/three-digit numbers with one/two decimal place/s by 10 and 100**	**Number unit 9:** Decimals: hundredths
2		
3		
4	• **Expressing percentages in decimal/ fractional form and vice versa** • **Finding a percentage of a set/quantity**	**Number unit 10:** Percentages
5		
6		**Data handling unit 2:** Spreadsheets and databases, language of probability

SHM 6   **Year:**

Class: **Group:**   **Block:**   **Date:**

WEEK	ORAL/MENTAL	MATHEMATICS TOPIC
1		
2		
3		
4		
5		
6		

Class: **Group:**   **Block:**   **Date:**

WEEK	ORAL/MENTAL	MATHEMATICS TOPIC
1		
2		
3		
4		
5		
6		

Delivering the Curriculum for Excellence

SHM 7   Year:

Class: **Group:**   **Block 1**   **Date:**

WEEK	ORAL/MENTAL	MATHEMATICS TOPIC
1	• **Adding/subtracting multiples of powers of 10 to/from numbers with up to nine digits**	**Number unit 1:** Place value
2	• **Rounding to the nearest 100/1000/ 1 000 000**	
3	• **Adding two-/three-digit numbers and four-digit multiples of 100** • **Adding rounded numbers to find approximate totals**	
4		**Number unit 2:** Addition
5		**Measure unit 1:** Length
6		**Shape unit 1:** 2D shape: rotational symmetry

Class: **Group:**   **Block 2**   **Date:**

WEEK	ORAL/MENTAL	MATHEMATICS TOPIC
1	• **Finding differences when certain four-digit numbers are involved** • **Subtracting two-/three-digit numbers**	**Number unit 3:** Subtraction
2	• **Subtracting rounded numbers to find approximate differences**	
3	• **Tables facts and multiplying by multiples of 10/100/1000** • **Multiplying a two-digit number by a single digit**	**Number Unit 4:** Multiplication
4	• **Multiplying rounded numbers to find approximate products** • **Multiplication strategies**	
5		**Measure unit 2:** Weight
6		**Measure unit 3:** Time: 24 hour times

SHM 7   Year:

Class: **Group:**   **Block 3**   **Date:**

WEEK	ORAL/MENTAL	MATHEMATICS TOPIC
1	• **Division facts and dividing by 100** • **Dividing certain numbers with up to four digits by a single digit**	**Number unit 5:** Division
2	• **Continuing number sequences** • **Using 'rules' to describe or generate number sequences**	
3	• **'Testing' numbers for exact divisibility by 2, 3, 4, 5, 6, 8, 9, 10, 25 and 100**	
4		**Number unit 6:** Number properties
5		**Data handling unit 1:** Surveys and databases
6		**Shape unit 2**: 2D shape: properties, patterns, construction

Class: **Group:**   **Block 4**   **Date:**

WEEK	ORAL/MENTAL	MATHEMATICS TOPIC
1	• **Forming equivalent fractions by multiplying/dividing** • **Finding a fraction of a set/quantity**	**Number unit 7:** Fractions
2		
3		
4		**Measure unit 4:** Area
5		
6		**Measure unit 5:** Time: durations, rate, speed **Shape unit 3**: 3D shape

Delivering the Curriculum for Excellence

SHM 7   Year:

Class: **Group:**   **Block 5**   **Date:**

WEEK	ORAL/MENTAL	MATHEMATICS TOPIC
1	• **Adding/subtracting one-/two-place decimals** • **Adding/subtracting rounded decimals to find approximate total/differences**	**Number unit 8:** Decimals: one- and two-place decimals
2	• **Multiplying/dividing by 2–9, 10 and 100** • **Multiplying/dividing rounded money amounts to find approximate products/quotients**	
3		
4		
5		**Measure unit 6:** Volume/Capacity: mixed measure
6		**Shape unit 4:** Position, movement and angle

Class: **Group:**   **Block 6**   **Date:**

WEEK	ORAL/MENTAL	MATHEMATICS TOPIC
1	• **Continuing sequences of three-place decimals** • **Adding/subtracting rounded decimals to find approximate totals/differences**	**Number unit 9:** Decimals: three-place decimals
2		
3		
4	• **Expressing percentages in decimal/ fractional form and vice versa** • **Finding a percentage of a set/quantity**	**Number unit 10:** Percentages
5		
6		**Data handling unit 2:** Interpreting graphs/data Probability: language, outcomes, experiments

SHM 7

Year:

Class: **Group:** **Block:** **Date:**

WEEK	ORAL/MENTAL	MATHEMATICS TOPIC
1		
2		
3		
4		
5		
6		

Class: **Group:** **Block:** **Date:**

WEEK	ORAL/MENTAL	MATHEMATICS TOPIC
1		
2		
3		
4		
5		
6		

Delivering the Curriculum for Excellence

Development charts

Problem-solving and enquiry activities are included throughout the materials. These activities challenge children to think, to question and to explain.

Data Handling	Number							
	Counting and place value	**Addition and Subtraction**	**Multiplication and Division**	**Money**	**Number properties**	**Fractions**	**Decimals**	**Percentages**
SHM 4								
- extracting information tally charts and bar charts - pictograms - Carroll diagrams - Venn diagrams	**Numbers to 10 000** - the sequences to 10 000 - place value, comparing and ordering - number names, ordinal numbers - estimating and rounding **Number properties** - odd and even numbers - counting on/back in 2s, 3s, 4s, 5s, 6s, 7s, 8s, 9s - rules for number sequences - multiples of 2, 3, 4, 5 and 10	**Addition to 1000** - addition to 100, mental strategies - two-digit numbers, bridging 100 - three-digit numbers, mental strategies - written methods of addition **Subtraction to 1000** - subtraction to 100, mental strategies - three-digit numbers, mental strategies - three-digit numbers, written procedures	**Multiplication** - table facts and multiplication by 10, 100 - multiplication beyond tables, mental strategies **Division** - by 2, 3, 4, 5 and 10 - by 8 - by 6 - by 9 - by 7 - consolidation - linking multiplication and division - two-digit numbers - remainders	**Money** - using £1 and £2 coins - using £5, £10 and £20 notes		**Fractions** - halves, quarters, tenths, thirds and fifths - sixths and eighths, equivalent fractions		
SHM 5								
- revises simple frequency axis scales - introduces bar line charts - extracting information - databases - spreadsheets - mean/average	**Numbers to 100 000** - number sequence to 10 000 - place value, comparing and ordering - number names, ordinal numbers	**Addition** - doubles and near doubles - involving three-digit numbers - involving four-digit numbers **Addition beyond 1000** - mental strategies **Subtraction** - mental subtraction, two-digit numbers - mental subtraction, three-digit numbers - written methods **Subtraction beyond 1000** - mental strategies	**Multiplication** - by 10, 100 - mental strategies - written strategies - using doubles - written methods **Division** - by 1–10, 100, 1000 - halving; linking multiplication and division - three-digit numbers; remainders	**Money** - using notes and coins - mental strategies	**Number properties** - number patterns and sequences - factors - square numbers - negative numbers	**Fractions** - halves, quarters, tenths, thirds and fifths - of a shape - equivalence - of a set/ quantity	**Decimals** - tenths	

Data Handling	Counting and place value	Addition and Subtraction	Multiplication and Division	Money	Number properties	Fractions	Decimals	Percentages
SHM 6 - introduces trends graphs - revises the range and mode - introduces the median and the mean - introduces compound bar graphs - introduces simple pie charts	**Numbers to millions** - number sequence to millions - place value - estimating and rounding	**Addition** - mental addition involving two-/three-digit numbers - involving numbers with up to four digits **Subtraction** - mental subtraction involving three-digit numbers - involving numbers with four or more digits	**Multiplication** - mental multiplication - written methods, calculator **Division** - mental division - written division, calculator		**Number properties** - number sequences and patterns - divisibility, multiples and factors, word formulae	**Fractions** - equivalence - fractions of a set/quantity; hundredths	**Decimals** - tenths - hundredths	**Percentages** - concept - links between fractions, decimals and percentages - percentages of shapes - of a set/quantity
SHM 7 - designing an efficient questionnaire - extracting information from a database - pie charts - bar charts with class intervals - continuous data - straight-line and curved-line graphs - the range, mode median and mean - probability: language, outcomes, experiments	**Place value** - numbers to hundreds of millions - place value - estimating and rounding	**Addition** - mental addition of numbers with up to five digits - mental/written addition of numbers with four or more digits **Subtraction** - mental subtraction involving numbers with up to five digits - mental/written subtraction involving numbers with four or more digits	**Multiplication** - mental multiplication - written multiplication, calculator **Division** - division by a single digit - division by two digits		**Number properties** - number types, sequences, patterns - formulae in word and symbols	**Fractions** - equivalence, thousandths - addition, subtraction, multiplication - ratio and proportion	**Decimals** - one- and two-place decimals: addition, subtraction and multiplication - one- and two-place decimals: division - three-place decimals	**Percentages** - links between notation for percentages, fractions and decimals - comparing and ordering - percentages of a set/quantity: using equivalent fractions and using a calculator

Delivering the Curriculum for Excellence

| Measure | | | | | Shape | | |
Length	Weight	Area	Volume/Capacity	Time	3D shape	2D shape	Position, movement and angle
SHM 4							
- revises estimating and measuring lengths to the nearest half metre - revises measuring in metres and centimetres - introduces measuring lengths to the nearest half centimetre - revises measuring using a tape measure - introduces measuring in metres and centimetres, for example, 2 m 30 cm	- revises the kilogram and half kilogram - introduces the gram - introduces the relationship 1 kg = 1000 g - deals with weighing in kilograms and half kilograms - provides estimating activities using kilograms and grams	- need for standard unit - square centimetre (cm²) - practice in measuring and drawing shapes involving $\frac{1}{4}$ square centimetres - investigates drawing shapes with the same area	- revises the litre and introduces the half litre - introduces millilitres - uses litres and millilitres in problem solving contexts	- the calendar - o'clock, quarter past, half past, quarter to - minutes past/to the hour - durations	- revises recognising and naming 3D shapes - deals with properties such as faces, edges, vertices - builds models of 3D shapes - builds 3D shapes from 2D pictures	- properties and patterns - line symmetry	- extends work on grid references - introduces co-ordinates - revises four compass directions N, S, E, W - introduces turns of 90°, 180°, 360° and 45° - introduces comparing and ordering angles
SHM 5							
- measuring in metres and centimetres - estimating and measuring lengths to the nearest $\frac{1}{4}$ metre - recording in decimal form - measuring and drawing lengths - choosing appropriate units and instruments - the kilometre - perimeter	- the kilogram/gram relationship - recording weights - estimating and weighing - reading scales - weighing and recording to the nearest 100 g	- finding area in square centimetres - formula for area of rectangles - finding approximate areas of irregular shapes - square metre	- litres, half litres and quarter litres - recording notation - scale reading - the cubic centimetre - finding volumes	- reading and writing times in 5 minute intervals - introduces reading and writing times in 1 minute intervals - durations	- recognising and naming 3D shapes - the octahedron - 3D shape properties	- properties - line symmetry - tiling and patterns	- co-ordinates - drawing shapes (including symmetrical) on a co-ordinate grid - moving shapes on a squared grid - following and describing pathways - the 8-point compass - acute and obtuse angles

	Length	Weight	Area	Volume/Capacity	Time	3D shape	2D shape	Position, movement and angle
SHM 6	- estimating and measuring - relationships: mm/cm, km/m - perimeters	- reading scales and recording - relationships: kg/g - practical contexts	- $A = l \times b$ - approximate area of squares, rectangles and irregular shapes - simple composite shapes - right-angled triangles	- relationships: ml/ℓ - tenth of a litre - reading scales - estimating and measuring - Imperial units	- analogue and digital displays - 12-hour notation and am/pm - 24-hour notation - duration - estimating and measuring	- nets of cube - identifying shapes from their nets - surface area - 2D representation of 3D shapes	- line symmetry - reflections where at least two sides are not parallel or perpendicular to line of symmetry - reflecting straight and curved lines - naming and using side and angle properties - equilateral, isosceles and right-angled triangles - adjacent sides - tiling patterns - circle designs - seven-piece tangram - pursuit curves	- 8-point compass and 12-point dial: multiples of 45° and 30° - bearings and three-figure notation - drawing rotations - estimating and measuring angles
SHM 7	- estimating and measuring - relationships: m/km, mm/cm - scale drawing - perimeters - practical contexts	- reading scales by estimating between marks - relationships: tonne/kg - practical contexts	- $A = l \times b$ - relationships: hectare/square metre/square kilometre - right-angled triangles - composite shapes - isosceles triangles	- relationships: ℓ/cl, cl/ml - tenth of a litre - reading scales - estimating and measuring - $V = l \times b \times h$ - Imperial units: volume, length, weight	- relationships among units of time - 24-hour notation - World time zones - durations - interpreting timetables - rate - speed	- relationships: faces/vertices/edges - identifying shapes from their nets - plan, front and side views - properties of edges: parallel/perpendicular	- line symmetry - rotational symmetry - naming and identifying quadrilaterals using side, angle and symmetry properties - diagonal properties of quadrilaterals - tiling patterns based on two shapes - constructing triangles	- co-ordinate in all four quadrants - acute, obtuse and reflex angles - estimating angles to the nearest 5° - measuring angles using a protractor - calculating the size of an angle - bearings and three-figure notation

Assessment record grid

SHM 5

Year: ☐ Class: ☐

Names	Numbers in 100 thousands — Check-up 1	Check-up 2	Check-up 3	Check-up 4	Addition — Check-up 5	Check-up 6	Subtraction — Check-up 7	Check-up 8	Multiplication — Check-up 9	Check-up 10	Check-up 11	Division — Check-up 12	Check-up 13	Check-up 14	Money — Check-up 15	Check-up 16	Fractions — Check-up 17	Decimals — Check-up 18	Check-up 19	Time — Check-up 20	Check-up 21	Topic Assessment — Numbers in 100 thousands	Addition	Subtraction	Addition and Subtraction beyond 1000	Multiplication	Division	Money	Fractions	Decimals	Level C test — Round-up 1	Round-up 2	Round-up 3

Assessment record grid

SHM 6

Year: [ ]　Class: [ ]

Names	Numbers in millions Check-up 1	Numbers in millions Check-up 2	Numbers in millions Check-up 3	Addition Check-up 4	Addition Check-up 5	Subtraction Check-up 6	Subtraction Check-up 7	Multiplication Check-up 8	Multiplication Check-up 9	Division Check-up 10	Division Check-up 11	Fractions Check-up 12	Fractions Check-up 13	Decimals Check-up 14	Decimals Check-up 15	Decimals Check-up 16	Decimals Check-up 17	Decimals Check-up 18	Decimals Check-up 19	Topic Assessment Addition	Topic Assessment Subtraction	Topic Assessment Multiplication	Topic Assessment Division	Topic Assessment Number properties	Topic Assessment Fractions	Topic Assessment Decimals	Topic Assessment Percentages	Topic Assessment Time	Topic Assessment Length	Topic Assessment Weight	Topic Assessment Area	Topic Assessment 2D Shape	Topic Assessment Position, Movement and Angle	Topic Assessment Data Handling	Round-up Level D

Delivering the Curriculum for Excellence

Assessment record grid

SHM 7

Names	Addition	Subtraction	Fraction		Decimals					Topic Assessment																		Round-up Level E
	Check-up 1	Check-up 2	Check-up 3	Check-up 4	Check-up 5	Check-up 6	Check-up 7	Check-up 8	Check-up 9	Place value	Addition	Subtraction	Multiplication	Division	Number properties	Fractions	Decimals	Percentages	Weight	Length	Time	Volume/Capacity	Area	2D shape	Position, Movement and Angle	Data Handling	

Level C Class record grid

Year: [ ] Class: [ ]

Names

Information handling

Collect

- **By obtaining information** for a task from a variety of given sources, including a simple questionnaire with yes/no questions.
- **By conducting a survey** which extends beyond the class.

Organise

- **By using a tally sheet** with grouped tallies.
- **By entering data in a table** using row and column headings.
- **By using a database** where the teacher defines the headings or fields.
- With the aid, where appropriate, of a computer package.

Display

- **By constructing a table or chart**.
- **By constructing a bar graph** with axes graduated in multiple units and discrete categories of information.
- With the aid, where appropriate, of a computer package.

Interpret

- **From displays and databases** by retrieving specific records and by identifying the most and least frequent items.
- With the aid, where appropriate, of a computer package.

Number, Money, Measure

Range and type of numbers

- **Work with** whole numbers up to 10 000 (count, order, read/write).
- **Work with** thirds, fifths, eighths, tenths and simple equivalences such as one half = two quarters (practical applications only).
- **Work with** decimals to two places when reading/recording money, and using calculator displays.

- **Money**
- **Use coins/notes** to £5 worth or more, including exchange.

Add

- Mentally for one digit to whole numbers up to three digits, beyond in some cases involving multiples of 10.
- Without a calculator for whole number with two digits added to three digits.
- With a calculator for three-digit whole numbers.
- In applications in number, measurement and money to £20.

Subtract

- Mentally for one digit from whole numbers up to three digits, beyond in some cases involving multiples of 10.
- Mentally for subtraction by 'adding on'.
- Without a calculator for whole numbers with two digits subtracted from three digits.
- With a calculator for three-digit whole numbers.
- In application in number, measurement and money to £20.

Multiply

- Mentally within the confines of all tables to 10.
- Mentally for any two- or three-digit whole number by 10.
- Without a calculator for two-digit whole numbers by any single- digit whole number.
- With a calculator for two- or three-digit whole numbers by a whole number with one or two digits.
- In applications in number, measurement and money to £20.

Delivering the Curriculum for Excellence

Level C Class record grid

Year: ☐ Class: ☐

Names

Number, Money, Measure

Divide
- Mentally within the confines of all tables to 10.
- Mentally for any two- or three-digit whole numbers by 10.
- Without a calculator for two-digit whole numbers by any single digit whole number.
- With a calculator for two- or three-digit whole numbers by a whole number with one or two digits.
- In applications in number, measurement and money to £20.

Round numbers
- **Round three-digit whole numbers** to the nearest 10.

Fractions, percentages and ratio
- **Find simple fractions** ($\frac{1}{3}$, $\frac{1}{5}$, $\frac{1}{10}$) of quantities involving one- or two-digit numbers.

Patterns and sequences
- **Work with patterns and sequences** within and among multiplication tables.

Functions and equations
- **Use a simple 'function machine' for operations** involving doubling, halving, adding and subtracting.

Measure and estimate
- **Measure in standard units:**
  - weight: accuracy extended to include 20 g weights, 1 kg=1000 g
  - volume: litre, $\frac{1}{2}$ litre, $\frac{1}{4}$ litre
  - area: shapes composed of rectangles/squares or irregular shapes using tiles or grids in square centimetres and metres.
- **Estimate length and height** in easily handled units: m, $\frac{1}{2}$ m, $\frac{1}{10}$ m, cm.
- **Select appropriate measuring devices and units** for length.
- **Read scales** on measuring devices to the nearest graduation where the value of the intermediate graduation may be deduced.
- **Realise that weight and area can be conserved** when shape changes.

Time
- Use 12-hour times for simple timetables.
- Conventions for recording time.
- Work with hours, minutes.
- Use calendars.

Shape, Position and Movement

Range of shapes
- **Collect, discuss, make and use 3D and 2D shapes**.
- Identify 2D shapes within 3D shapes.
- Draw circles using a variety of methods.
- Recognise 3D shapes from 2D drawings.

Position and movement
- Describe the main features of a familiar journey or route.
- Create paths on squared paper described by instructions such as 'forward 5, right 90, forward 7, left 90'.

Symmetry
- Find lines of symmetry of shapes drawn on squared grids.
- Complete the missing half of a simple symmetrical shape or pattern on a squared grid.

Angles
- Know that a right angle is 90°.
- Use 'right, acute, obtuse' to describe angles.
- Know that a straight angle is 180°.

Level D Class record grid

Year: [ ] Class: [ ]

Names

Information handling

Collect
- **By selecting sources of information** for tasks, including a questionnaire which allows several responses to each question.

Organise
- **By using diagrams or tables.**
- **By using a database or spreadsheet table** with up to three fields defined by pupils.
- With the aid, where appropriate, of a computer package.

Displays
- **By constructing graphs (bar, line, frequency, polygon) and pie charts.**
- Involving simple fractions or decimals.
- Involving continuous data which has been grouped.
- With the aid, where appropriate, of a computer package.

Interpret
- **From a range of displays and databases** by retrieving information subject to one condition.

Number, Money, Measure

Range and type of numbers
- **Work with** whole numbers up to 100 000 (count, order, read/write).
- **Work with** whole numbers to a million (read/write only).
- **Work with** fractions (all previous plus twentieths, fiftieths, hundredths) and equivalences among these and decimals (in applications).
- **Work with** percentages, decimals to two places and equivalences among these in applications in money and measurement.

Money
- **Use all UK coins/notes** to £20 worth or more, including exchange.

Add
- Mentally for two-digit whole numbers, beyond in some cases, involving multiples of 10 or 100.
- Without a calculator, for four digits with at most two decimal places (easy examples only).
- With a calculator, for four digits with at most two decimal places.
- In applications in number, measurement and money.

Subtract
- Mentally for two-digit whole numbers, beyond in some cases, involving multiples of 10 or 100.
- Without a calculator, for four digits with at most two decimal places (easy examples only).
- With a calculator, for four digits with at most two decimal places.
- In applications in number, measurement and money.

Multiply
- Mentally for whole numbers by single digits: easy examples only.
- Mentally for four digit numbers including decimals by 10 or 100.
- Without a calculator for four digits with at most two decimal places by a whole number with two digits.
- In applications in number, measurement and money.

Divide
- Mentally for whole numbers by single digits: easy examples only.
- Mentally for four digit numbers including decimals by 10 or 100.
- Without a calculator for four digits with at most two decimal places by a whole number with two digits.
- In applications in number, measurement and money.

Delivering the Curriculum for Excellence

Level D Class record grid

Year: ☐ Class: ☐

<table>
<tr><td></td><td rowspan="99">Names</td><td></td><td></td><td></td><td></td><td></td><td></td><td></td><td></td><td></td><td></td></tr>
<tr><td colspan="1">Number, Money, Measure</td><td></td><td></td><td></td><td></td><td></td><td></td><td></td><td></td><td></td><td></td></tr>
</table>

Number, Money, Measure

Round numbers
- **Round any number** to the nearest appropriate whole number, 10 or 100.

Fractions, percentages and ratio
- **Work with fractions and percentages**: find simple fractions ($\frac{1}{7}$, $\frac{3}{4}$, $\frac{3}{5}$, $\frac{60}{100}$) of quantities involving at most four digits (easy examples only).

Patterns and sequences
- **Continue and describe more complex sequences.**

Functions and equations
- **Recognise and explain simple relationships** between two sets of numbers of objects.

Measure and estimate
- **Measure in standard units:**
  - length: small lengths in millimetres; large lengths like buildings in metres
  - weight: extended range of articles, for example own weight
  - volume: accuracy extended to small containers in millimetres; 1 ℓ = 1000 ml
  - area: right-angled triangles on cm squared grids
  - temperature.
- **Estimate** small weights, small areas, small volumes in easily handled standard units.
- **Recognise** when kilometres are appropriate.
- **Select appropriate measuring devices and units** for weight.
- **Be aware of common Imperial units** in appropriate practical applications.

Time
- Use 24-hour times and equate with 12-hour times.
- Calculate duration in hours/minutes, mentally if possible.
- Time activities in seconds with a stopwatch.
- Calculate speeds (practical activities only).

Perimeter, formulae, scales
- **Calculate perimeter** of simple straight-sided shapes by adding lengths.

Shape, Position and Movement

Range of shapes
Collect, discuss, make and use 3D and 2D shapes
- Discuss 3D and 2D shapes referring to faces, edges, vertices, diagonals, sides, angles.
- Recognise pentagon, hexagon.
- Identify and name equilateral and isosceles triangles.
- Extend shape vocabulary to radius, diameter, circumference.
- Create or copy a tiling using a shape template.
- Make 3D models, solid or skeletal, including using nets: cube and cuboid only.
- Use the rigidity property of triangles in model-making.

Position and movement
- Give directions for a route or journey.
- Use an 8-point compass rose.
- Use a co-ordinate system to locate a point on a grid.
- Create patterns by rotating a shape.

Symmetry
- Identify and draw lines of symmetry, generally up to four.
- Create symmetrical shapes.

Angles
- Draw, copy and measure angles accurately within 5 degrees.
- Use standard notation, 060°, 150°, 300°, to express bearings.

Level E Class record grid

Year: ☐ Class: ☐

Names

Information handling

Collect
- **By selecting sources of information** for tasks, including:
  - practical experiments
  - surveys using questionnaires
  - sampling using a simple strategy.

Organise
- **By designing and using diagrams and tables.**
- **By designing and using a database or spreadsheet** with fields defined by pupils. With the aid, where appropriate, of a computer package.

Display
- **By constructing straight line and curved graphs for continuous data** where there is a relationship such as direct proportion – travel, temperature, growth graphs.
- **By constructing pie charts of data** expressed in percentages. With the aid, where appropriate, of a computer package**.**

Interpret
- **From an extended range of displays** (diagrams, tables, graphs, pie charts) **and databases**, retrieving information subject to more than one condition. With the aid, where appropriate, of a computer package, involving the use of logical operators (and; or; not).
- **By describing the main features of a graph** so as to show an awareness of the significance of the information.
- **By calculating the average (mean)** to compare sets of data.

Number, Money, Measure

Range and type of numbers
- **Work with:**
  - all widely used fractions and equivalence among these and decimals (in applications)
  - decimals to three places (practical applications in measurement).

Money
- **Use relationships between currencies** to do simple calculations.

Add
- Mentally for 2-digit numbers including decimals.
- Without a calculator for four digits with at most two decimal places.
- With a calculator for any number of digits with at most three decimal places in applications in number, measurement and money.

Subtract
- Mentally for 2-digit numbers including decimals.
- Without a calculator for four digits with at most two decimal places.
- With a calculator for any number of digits with at most three decimal places in application in number, measurement and money.

Multiply
- Mentally for any whole number by a multiple of 10 or 100 (such as 20 or 200).
- Mentally for any numbers including decimals by 10, 100, 1000.
- Without a calculator for four digits with at most two decimal places in the answer in applications in number, measurement and money.

Delivering the Curriculum for Excellence

Level E Class record grid

Year: [ ] Class: [ ]

Names

Number, Money, Measure										

Divide
- Mentally for any whole number by a multiple of 10 or 100 (such as 20 or 200).
- Mentally for any numbers including decimals by 10, 100 or 1000.
- Without a calculator for four digits with at most two decimal places by a single digit.
- With a calculator for any pair of numbers but at most three decimal places in the answer.
- In applications in number, measurement and money.

Add and subtract
- **Work with:**
  - negative number (eg temperature)
- **Add and subtract:**
  - positive and negative numbers in application such as rise in temperature.

Round numbers
- **Round any number** to one decimal place.

Fractions, percentages and ratios
- **Work with fractions:**
  - mentally find widely-used percentages of whole number quantities
  - with a calculator find a percentage of a quantity
  - without a calculator as previously defined.
- **Find ratios between quantities.**
- **Use simple unitary ratio.**

Patterns and sequences
- **Continue and describe sequences:**
  - involving square and triangular numbers
  - finding specified items in sequences
  - prime numbers.

Functions and equations
- **Use a 'function machine' in reverse for inverse operations.**
- **Solve simple equations and inequations.**
- **Use notation** to describe general relationships between two sets of numbers.
- **Use and device simple rules.**

Measure and estimate
- **Measure and draw using standard units:**
  - accuracy and device as appropriate to the application.
- **Estimate measurements:**
  - areas in square metres
  - small lengths in millimetres
  - larger lengths in metres.
- **Work with square kilometre, hectare, tonne** when appropriate.
- **Read scales** on measuring devices, including estimating between graduations.

Time
- **Time activities** with a digital stopwatch in seconds, tenths, hundredths.

Perimeter, formulae, scales
- **Calculate using rules:**
  - areas of rectangles and squares
  - volume of cuboids and cubes.
- **Use scales** such as 1 cm to 1, 2, 5 or 10 m: or represented by a ratio such as 1:100 to interpret or draw maps, plans, diagrams, or to make models.

Level E Class record grid

Year: [ ] Class: [ ]

Names

Shape, Position and Movement

Range of shapes
- **Use properties of 2D and 3D shapes:**
  - discuss the side, angle, diagonal of quadrilaterals: square, rectangle, rhombus, parallelogram, kite, trapezium
  - define and classify quadrilaterals
  - relate diameter and circumference (practical examples only)
  - make 3D models, solid or skeletal, including using nets: triangular prism, pyramid, tetrahedron.
- **Draw triangles:**
  - given three sides.

Position and movement
- **Discuss position and movement:**
  - use bearings and distances to produce accurate scale drawings of routes
  - use co-ordinates in all four quadrants to plot position
  - calculate distances along grid lines.

Symmetry
- **Work with symmetry:**
  - determine whether or not shapes have rotational symmetry
  - move a tile of a shape on a squared grid in order to translate, reflect or rotate the shape.

Angle
- **Angles:**
  - use 'reflex' to describe angles
  - know the sum of the angles of a triangle is two right angles.

Delivering the Curriculum for Excellence

Record of work: SHM 5   Name: _____   Year: ☐   Class: ☐

Numbers in 100 thousands

Number sequence to 100 thousands	TB 1	TB 2	HA 1	CU 1	

Place, value, comparing and ordering	TB 3	TB 4	HA 2	CU 2	EX 1

Number names, ordinal numbers	TB 5	TB 6	CU 3

Estimating and rounding	TB 7	TB 8	TB 9	CU 4	TA 1a
	TA 1b				

Addition

Doubles and near doubles	TB 10	TB 11	TB 12	HA 3	CU 5

Addition involving three-digit numbers	TB 13	TB 14	TB 15	TB 16	TB 17
	HA 4	CU 6			

Addition involving four-digit numbers	TB 18	TB 19	TB 20	TB 21	EX 2
	TA 2a	TA 2b			

Subtraction

Mental subtraction involving two-digit numbers	TB 22	TB 23	TB 24	HA 5	CU 7

Mental subtraction involving three-digit numbers	TB 25	TB 26	TB 27	HA 6	CU 8

Written methods of subtraction	TB 28	TB 29	EX 3	TA 3a	TA 3b

Addition and Subtraction beyond 1000

Addition beyond 1000: mental strategies	TB 30	TB 31

Subtraction beyond 1000: mental strategies	TB 32	TB 33	TB 34	TB 35	EX 4
	EX 5	EX 6	TA 4a	TA 4b	

Record of work: SHM 5   Name: [_____]   Year: [__]   Class: [__]

Multiplication

Multiplication by 10, 100	TB 36	TB 37	HA 7	CU 9

Mental and written strategies	TB 38	TB 39	TB 40	TB 41	TB 42
	TB 43	HA 8	HA 9	CU 10	CU 11

Using doubles	TB 44

Written methods of multiplication	TB 45	TB 46	TB 47	EX 7	TA 5a
	TA 5b				

Division

Dividing by 2–10, 100, 1000	TB 48	TB 49	TB 50	TB 51	HA 10
	CU 12				

Halving; linking multiplication and division	TB 52	TB 53	CU 13

Dividing 3-digit numbers; remainders	TB 54	TB 55	CU 14	EX 8	EX 9
	TA 6a	TA 6b			

Money

Amounts using notes and coins	TB 56	TB 57	HA 11	CU 15

Using mental strategies	TB 58	TB 59	TB 60	TB 61	HA 12
	HA 13	CU 16			

Fractions

Halves, quarters, tenths, thirds and fifths	TB 62

Fraction of a shape, equivalence	TB 63	TB 64	TB 65	HA 14	CU 17

Fraction of a set/ quantity	TB 66	HA 15	EX 10	TA 8

Record of work: SHM 5   Name: [ ]   Year: [ ]   Class: [ ]

Decimals

Tenths	TB 67	TB 68	TB 69	TB 70	TB 71
	TB 72	TB 73	HA 16	HA 17	CU 18
	CU 19	EX 11	EX 12	EX 13	TA 9a
	TA 9b				

Number properties

Number properties	TB 74	TB 75	TB 76	EX 14	EX 15
	EX 16	EX 17			

Measure

Length	TB 77	TB 78	TB 79	TB 80	TB 81
	TB 82	EX 18	EX 19		

Weight	TB 83	TB 84	TB 85	TB 86	TB 87
	TB 88				

Volume/Capacity	TB 89	TB 90	TB 91	TB 92

Area	TB 93	TB 94	TB 95	EX 20

Time: reading and writing times

	TB 96	HA 18	CU 20

Time: durations

	TB 97	TB 98	TB 99	TB 100	TB 101
	CU 21				

Shape

2D shape: properties	TB 102	TB 103	TB 104

2D shape: line symmetry, tiling and patterns	TB 105	TB 106	TB 107	TB 108	TB 109

3D shape	TB 110

Position, movement and angle	TB 111	TB 112	TB 113	TB 114	TB 115
	TB 116	TB 117	TB 118		

Data handling

Using the data handling process	TB 119	TB 120	TB 121	TB 122	TB 123
	EX 21	EX 22			

Record of work: SHM 6   Name: _____   Year: ☐ Class: ☐

Numbers to millions

Number sequence to millions	TB 1	HA 1	CU 1		

Place value	TB 2	TB 3	TB 4	TB 5	TB 6
	HA 2	HA 3	HA 4	CU 2	EX 1
	EX 2				

Estimating and rounding	TB 7	TB 8	CU 3

Addition

Mental addition involving two-/ three- digit numbers	TB 9	TB 10	TB 11	TB 12	TB 13
	HA 5	CU 4			

Addition involving numbers with up to four digits	TB 14	TB 15	TB 16	TB 17	TB 18
	HA 6	CU 5	TA 1a	TA 1b	

Number properties

Number sequences and patterns	TB 37	TB 38	TB 39	HA 14	EX 4
	EX 5	EX 6			

Divisibility, multiples and factors, word formulae	TB 40	TB 41	TB 42	TB 43	TB 44
	TB 45	TB 46	TB 47	TA 5a	TA 5b

Fractions

Equivalence	TB 48	TB 49	TB 50	HA 15	CU 12
	EX 10				

Fraction of a set/quantity; hundredths	TB 51	TB 52	TB 53	HA 16	CU 13
	TA 6a	TA 6b			

Subtraction

Mental subtraction involving three-digit numbers	TB 19	TB 20	TB 21	HA 7	CU 6

Subtraction involving numbers with four or more digits	TB 22	TB 23	TB 24	HA 8	CU 7
	EXT 3	HA 2a	TA 2b		

Multiplication

Mental multiplication	TB 25	TB 26	TB 27	TB 28	TB 29
	HA 9	HA 10	HA 11	CU 8	CU 9

Written methods, calculator	TB 30	TB 31	TB 32	TA 3a	TA 3b

Division

Mental division	TB 33	TB 34	HA 12	HA 13	CU 10

Written division, calculator	TB 35	TB 36	CU 11	TA 4a	TA 4b

Decimals

Tenths	TB 54	TB 55	TB 56	TB 57	TB 58
	TB 59	TB 60	HA 17	HA 18	CU 14
	CU 15	CU 16			

Hundredths	TB 61	TB 62	TB 63	TB 64	TB 65
	TB 66	TB 67	TB 68	HA 19	HA 20
	HA 21	EX 8	EX 9	TA 7a	TA 7b

Percentages

Percentages	TB 69	TB 70	TB 71	TB 72	HA 22
	EX 11	TA 8a	TA 8b		

Delivering the Curriculum for Excellence

Record of work: SHM 6 Name: [____] Year: [ ] Class: [ ]

Measure

Time: reading and writing times	TB 73	TB 74	HA 23		

Time: durations, seconds	TB 75	TB 76	TB 77	TB 78	EX 12
	EX 13	TA 9			

Length	TB 79	TB 80	TB 81	TB 82	TB 83
	TB 84	TA 10			

Weight	TB 85	TB 86	TA 11

Volume/Capacity	TB 87	TB 88	TB 89	TB 90

Mixed measure	TB 91	TB 92

Area	TB 93	TB 94	TB 95	TB 96	TB 97
	EX 19				

Shape

2D shape: line symmetry	TB 98

2D shape properties, puzzles and patterns	TB 99	TB 100	TB 101	TB 102	TB 103
	TB 104	EX 14	EX 15	EX 16	EX 17
	EX 18	TA 13a	TA 13b		

3D shape	TB 105	TB 106	TB 107	TB 108

Position, movement and angle	TB 109	TB 110	TB 111	TB 112	TA 14a
	TA 14b				

Data handling

Interpreting graphs/data	TB 113	TB 114	TB 115	TB 116	TB 117
	EX 20	EX 21			

Bar charts with class intervals	TB 118	EX 22

Spreadsheets and databases	TB 119	TB 120	TB 121	TB 122

Language of probability	TB 123	TA 15

Record of work: SHM 7   Name: _____   Year: ☐ Class: ☐

Place value

Numbers to hundreds	TB 1	TB 2	TB 3	TB 4	TB 5
	TB 6	HA 1	HA 2	EX 1	TA 1a
	TA 1b				

Addition

Mental addition of numbers with up to five digits	TB 7	TB 8	TB 9	TB 10	HA 3
	CU 1				

Mental/written addition involving numbers with four or more digits	TB 11	TB 12	EX 3	TA 2a	TA 2b

Subtraction

Mental subtraction involving numbers with up to five digits	TB 13	TB 14	HA 4	CU 2

Mental/written subtraction involving numbers with four or more digits	TB 15	TB 16	TB 17	TB 18	EX 3
	TA 3a	TA 3b			

Multiplication

Mental multiplication	TB 19	TB 20	TB 21	HA 5

Written multiplication, calculator	TB 22	TB 23	TB 24	EX 4	EX 22
	TA 4a	TA 4b			

Division

Division by a single digit	TB 25	TB 26	TB 27	HA 6

Division by two digits	TB 28	TB 29	TB 30	EX 5	TA 5a
	TA 5b	TA 5c			

Number properties

Number types, sequences, patterns	TB 31	TB 32	TB 33	TB 34	TB 35
	HA 7	HA 8			

Formulae in words and symbols	TB 36	TB 37	TA 6a	TA 6b

Fractions

Equivalence, thousandths	TB 38	TB 39	HA 9	CU 3

Addition, subtraction, multiplication	TB 40	TB 41	TB 42	TB 43	TB 44
	CU 4	EX 6			

Ratio and proportion	TB 45	TB 46	EX 7	TA 7a	TA 7b

Decimals

One- and two-place decimals; addition, subtraction and multiplication	TB 47	TB 48	TB 49	TB 50	TB 51
	TB 52	HA 10	HA 11	CU 5	CU 6

One- and two-place decimals: division	TB 53	TB 54	TB 55	TB 56	TB 57
	TB 58	TB 59	HA 12	CU 7	CU 8

Three-place decimals	TB 60	TB 61	TB 62	TB 63	HA 13
	CU 9	EX 8	EX 9	TA 8a	TA 8b
	TA 8c				

Percentages

Percentages	TB 64	TB 65	TB 66	TB 67	TB 68
	HA 14	EX 10	EX 11	TA 9a	TA 9b

Problem solving

Problem solving and enquiry	TB 69	TB 70	TB 71	TB 72	TB 73

Delivering the Curriculum for Excellence

Record of work: SHM 7　Name: [_____]　Year: [ ]　Class: [ ]

Measure

Weight	TB 74	TB 75	TB 76	HA 15	TA 10

Length	TB 77	TB 78	TB 79	TB 80	TB 81

	TB 82	TA 11

Time: 24-hour time	TB 83	TB 84	HA 16

Time: durations, rate, speed	TB 85	TB 86	TB 87	TB 88	TB 89

	TB 90	HA 17	EX 12	EX 13	TA 12a

	TA 12b

Volume/Capacity: volume/capacity	TB 91	TB 92	TB 93	TB 94	TA 13

Volume/Capacity: mixed measure	TB 95	TB 96

Area	TB 97	TB 98	TB 99	EX 14	TA 14

Shape

3D shape	TB 100	TB 101	TB 102	EX 15

2D shape: rotational symmetry	TB 103	TB 104

2D shape properties, patterns, construction	TB 105	TB 106	EX 16	EX 17	TA 15

Position, movement and angle	TB 107	TB 108	TB 109	TB 110	TB 111

	TB 112	TB 113	EX 18	EX 19	TA 16a

	TA 16b

Data handling

Surveys and databases	TB 114	TB 115	TB 116

Interpreting graphs/data	TB 117	TB 118	TB 119	TB 220	TB 221

	TA 17a	TA 17b

Probability: language, outcomes, experiments	TB 122	TB 123	EX 20	EX 21